Oil and Gas Forecasting

INTERNATIONAL ASSOCIATION FOR MATHEMATICAL GEOLOGY
STUDIES IN MATHEMATICAL GEOLOGY

1. William B. Size (ed.):
Use and Abuse of Statistical Methods in the Earth Sciences

2. Lawrence J. Drew:
Oil and Gas Forecasting: Reflections of a Petroleum Geologist

Oil and Gas Forecasting

Reflections of a Petroleum Geologist

Lawrence J. Drew

Branch of Resource Analysis
U.S. Geological Survey, Reston, Virginia

New York Oxford
OXFORD UNIVERSITY PRESS
1990

Oxford University Press

Oxford New York Toronto
Delhi Bombay Calcutta Madras Karachi
Petaling Jaya Singapore Hong Kong Tokyo
Nairobi Dar es Salaam Cape Town
Melbourne Auckland

and associated companies in
Berlin Ibadan

Published by Oxford University Press, Inc.,
200 Madison Avenue, New York, New York 10016

Library of Congress Cataloging-in-Publication Data
Drew, Lawrence J.
Oil and gas forecasting : reflections of a petroleum geologist /
Lawrence J. Drew.
p. cm. — (Studies in mathematical geology : 2)
Bibliography: p. Includes index.
ISBN 0-19-506170-5
1. Petroleum—Reserves—Forecasting.
2. Gas, Natural—Reserves—Forecasting.
I. Title. II. Series.
TN871.D715 1990
622'.1828—dc20 89-8633 CIP

2 4 6 8 9 7 5 3 1

Printed in the United States of America
on acid-free paper

Foreword to the Series

This series of studies in mathematical geology provides contributions from the geomathematical community on topics of special interest in the earth sciences. As far as possible, each volume in the series will be self-contained and will deal with a specific technique of analysis. For the most part, the results of research will be emphasized. An important part of these studies will be an evaluation of the adequacy and the appropriateness of present geomathematical and geostatistical applications. It is hoped the volumes in this series will become valuable working and research tools in all facets of geology. Each volume will be issued under the auspices of the International Association for Mathematical Geology.

Richard B. McCammon
U.S. Geological Survey
Reston, Virginia

Preface

My purpose in writing this book is to relate the events and expose the human drama that unfolded as advances were made in the field of petroleum geology. My recollections are based on more than twenty years as a witness and a participant in this field. The positions staked out by the major figures are mentioned along with bits of useful gossip and anecdotes. The names of most of the people involved are disclosed, although partial disguises have been used where discretion dictates. In a broad sense, the book is about petroleum resource assessment; in a narrow sense, it is about forecasting oil and gas discovery rates and the associated task of determining the distributional form of oil and gas field size distributions. More recently, I have tried my hand at mineral resource assessment. The last chapter chronicles some of the problems encountered and the attempts made at their solution.

Reston, Virginia
January 1990

L. J. D.

Acknowledgments

I wish to thank all those who contributed directly or indirectly to the preparation of this book. I thank the series editors, Richard B. McCammon, C. John Mann, and Thomas A. Jones; John H. DeYoung, Jr., who read the first draft of the manuscript; David H. Root for his good council and the many pieces of historical information that he contributed; John H. Schuenemeyer and Emil D. Attanasi for the privilege of working with them; and everyone else who helped. I wish to extend a special thanks to Janet Somerville Sachs for her editorial expertise and unflagging goodwill.

Contents

Oil and Gas Forecasting

1

Reflections of a Petroleum Geologist

The earliest event in my life that I can connect to my interest in forecasting the occurrence of natural resources was when I filled out an order form for a book that promised to tell how to discover uranium deposits on the Colorado Plateau. Where the order form came from is a mystery now, but I can clearly remember buying a postal money order for $4.16, sending it off to the bookseller, and waiting each day for the mailman to bring my treasure. After what seemed an eternity, the book finally came, and I read it as best as my junior high school education would allow. I had an immediate interest in how geiger counters and geochemical surveys were used to discover uranium deposits and little interest in how to operate a uranium mine for profit after its discovery.

Sometime in the early 1960s when I was in graduate school, I remember my mother telling me that she and my father had spent hours wondering what had made me go into geology when neither of them had ever mentioned the idea at any time while I was growing up. Although the reason escapes me as to why I was originally interested in geology, I know that my involvement in forecasting oil and gas field discovery rates was a consequence of my tutelage under a Welshman by the name of John Cedric Griffiths.

Professor Griffiths was chairman of the department of mineralogy and petrology in the College of the Mineral Industries at the Pennsylvania State University, the department I entered to start my graduate studies. Our first encounter was a sort of interview in which I was assigned my first-quarter courses. As men in his position often do, Professor Griffiths asked me just what

I wanted to do in graduate school. I think I said something about wanting to study the origin of mineral deposits. I did not catch on until years later that he had no intention of letting me study the origin of any geologic entity about which I could be no more specific than "the origin of mineral deposits." Perhaps if I had been more specific in this initial interview, I might have gotten away. I was not, and my reformation was about to begin. This man, who insisted on being addressed simply as "Griff," had assigned himself the task of stripping away all the deterministic dogma that had been inculcated in me during my previous 16 years in school. The essence of this task was to make me think as an experimental design statistician thinks. Griff's model for his students came out of the heart of philosophy of statistical empiricism.

I soon learned that he enjoyed conflict and was skilled in the use of the English language. He wrote very well, but, most of all, he enjoyed speaking in front of an audience. Many geologists have told stories about how they fared when they went up against Griff in public. Michel David, who is a bright young man and a master of geostatistics, told me a story that I feel typifies Griff. A few years ago, Michel walked into my office at the U.S. Geological Survey (USGS) and, in the course of conversation, said that he had recently debated my mentor. I could not resist asking him how it went. He said that Griff had beaten him twice on two consecutive days. On the first day, Michel had taken one side of the argument, and Griff had taken the other. He said that he lost the argument that day because Griff had better technical material. So he decided to take Griff's position on the second day. Griff, sensing the challenge of the young man, willingly took Michel's position from the previous day. It was the result of the debate on the second day that really impressed Michel. Although he lost again, he spoke of his defeat as if it were an honor. I told him that, in my experience, the closest Griff ever came to being bested in argument was in one of his classes when three of us ganged up on him. We did not win, but we had Griff on the ropes for a few minutes. The issue was whether it was possible to extract causality from a data matrix by using no paradigm for what cause and effect must look like. We had Griff on the defensive because, between the three of us, we knew a little more linear algebra than he did. He slipped from our grasp when he changed the premise of the argument. We saw it happening but were powerless to do anything about it. We consoled ourselves by concluding that, in addition to being Welsh, he had been trained at the Royal College of Science, London. We were American schoolboys who could read English and sort of write the stuff, but, when it came to speaking it in public, we had a lot to learn.

Like no other man I have known, Griff could say he had read the scientific literature on sedimentary petrography and petrology; not only had he read the literature in English, but also in French. His ability to read French went far beyond the translation of words. I remember he said that he knew how Cailleux approached the science, what he saw, and how he thought. He often spoke of what it meant to be from Gaul and said pointedly that I must come to understand such things. Griff was always aware of the set of presuppositions that a scientist used to make his observations and to interpret their meaning. Near the end of my graduate training, Eugene Williams, another of my professors and one of Griff's antagonists, tried to caution me before it was too late about being brainwashed by the fiery preaching of a Welshman and walking down his uncharted paths. Williams claimed that he should know such things because he came from the same stock and knew a sermon when he heard one.

At that time and, for the most part, today, Griff made good sense. I had to agree with him when he pointed out that the application of mathematical and statistical principles to geology was really just beginning. In 1962, computers were new on campus. Griff made the telling point that we were no longer bound to mechanical calculating machines. The computer could do a discriminate analysis in minutes rather than days, and it did not make mistakes! He argued that we could leave the dark ages and make geology a real empirical science, maybe even a statistical science. I remember Griff drawing a Venn diagram on a piece of lined yellow paper to emphasize a point he was making. I looked down in disbelief at what was nothing more than an unintelligible hieroglyph. I wondered what I had gotten myself into as he chomped down on his nickel cigar, pointed at the Venn diagram, and said this was the way to the future. The future, as he saw it, was in statistics, mathematics, operations research, and the computer.

Above all, Griff believed in the computer. The computer made an enormous impact because, before it came on the scene, applied statistics had been a tedious business. It took a special kind of concentration and care to pound an old Monroe or Frieden mechanical calculator for hours on end. You always had to be mindful of the checksums and the rounding error and the headaches and especially the eye strain which had to be the worst part of the activity. It soon became obvious that he saw more in the computer than a simple labor-saving device. I was never sure, but I believe that it had something to do with the fact that, to him, the computer was totally objective. Griff even talked about training a computer to think rationally for itself.

The recipe for my conversion away from determinism and toward statistical empiricism required taking more different types of courses than I had ever imagined. There were courses in probability theory and mathematical statistics, and, for some of my fellow students, it was off to the agronomy department for courses in experimental design. Griff sent me to the psychology department to try out their series on analysis of variance instead—better I had taken the longer walk to the agronomy building! One of my classmates went off to the industrial engineering department to learn operations research techniques. I remember one math course in stochastic processes that I took with a bunch of electrical engineers that made me very respectful of that field. Sometimes we felt rather lonely in these classes. There just were not very many students in the entire College of the Mineral Industries who took a substantial portion of their course work in agronomy, psychology, and mathematics and who spent their evenings at the computer center. A course or two in the chemistry department was common enough, but everyone could see that Griff had something different in mind for his students.

Sprinkled among the classes we took were Griff's own. These were a series of four classes that could be taken over a 2-year period. The class size was small, 6 to 10 students, most of whom either were his students or were trying to learn applied statistics from someone who had gained a reputation for knowing how to do it. At times, the subject matter of these courses was not laid out in an order with which the conventional mid–twentieth-century mind could grapple easily. I remember one course opening with a lecture on the structure of a sensor. The stuff of the lecture was taken from a book on servomechanisms. In the middle of the lecture, Griff looked us over slowly and then turned and drew a potato on the board. The potato had cilia on it. He turned back to the class and said we were all like this potato—when we sensed something, we wiggled our arms. To us, this was a shocking, outrageous statement, but Griff intended to do many thngs with this remark. First and foremost, it was an iconoclastic bash at our traditional educational system. Griff also was saying that the whole mechanism by which we thought had run amuck because we did not realize that we were part of the system that was sensing the input data and also making the judgments. Without knowing, we assumed that we could reach the truth by standing back from it all and making independent analytical determinations.

Griff did not stop there because he felt that we as students had some nerve to assume, without question, that we knew the characteristics of truth. He would pound away at the distinction between precision and accuracy. He was sure we could be precise, but he knew very well that we did not know accuracy from a hole in the ground. Accuracy, after all, is the truth, and how did one know the truth? The correct answer, according to Griff, was that you assumed the truth. The verbal blasts came over and over again, "You do not know what question to ask until you know the answer or at least the form of the answer." Precision was only a measure of reproducibility. He made the point over and over again that, if we could so define precision, we could make sure we got reproducibility. Griff knew that, sooner or later, we had to ask ourselves, "What is accuracy?" and when we did, we would probably not like our answers. He knew that we were on the path he had set out for us by the questions we asked, and he waited patiently for our arrival at the destination he had targeted for us.

He was making the same point when he required us to read a book entitled *Flatland.* The message was that you could not be too sure of your deductions because you really did not know the dimensionality of the system you were in at any moment. You could use all your cognitive powers in two-dimensional thought, but, if the problem was really in three or more dimensions, you could reason incorrectly and never know that you were dead wrong. There was one lecture based solely on the Sunday magazine section of the *New York Times.* When this occurred, the students in the class could only conclude that he had forgotten to prepare a lecture. To the unindoctrinated, going to Griff's lectures was a mad scramble over a bunch of verbal high hurdles; to the *cognoscenti,* however, it became the narrow path.

Lectures on scale theory were also a staple. It was here that you learned that zero had many meanings and only rarely was it coincident with the notion of nothingness or emptiness. It was pointed out that only in the roughest sense could a determination of 20 parts per million copper in a geochemical exploration sample be interpreted as being a number twice as large as a determination of 10 parts per million.

The types of statistical tests you could use were linked to the scale of the data you collected. Griff would ask what kind of data we had collected and onto what measurement scale it had been mapped. Griff often used exaggeration to make a lasting impression. He once said that, in sedimentology, sieving sediment to determine grain size was a crude procedure no better than trying to measure the size of a desk by trying to shove it through a door. I remember the interrogation that came with this example: "Is that how you would measure a desk, Mr. Drew?" I answered that I would use a tape measure. "Why would you use a tape measure?" came the question. I connected the concept of a desk with ratio scale measurement versus measurement by sieving, which may be an ordinal scale at best. With this I was off the hook, yet, when Griff looked down at me, he was not quite sure what had been learned by his pup.

After these shock treatments had destroyed the old deterministic ways of thinking, it was time for the restructuring of the students' minds. This was done by providing us with a new building block—the concept of the random sample. Griff would emphasize the point that when you have a random sample, you have something very special—accuracy is now within your grasp! Well, almost. It is possible to be accurate within the concept of a probabilistic statement. This is, after all, the very best a human being can do in the search for objectivity and truth. The random sample was offered as our exit visa to freedom from the troubled deterministic world. We could now put bias and prejudice behind us and be objective. With parametric statistical procedures applied to the data collected by means of random sampling, we could ride on the crest of a wave headed for the future. We had broken the shackles of determinism. Griff liked to say that we were finally cooking with gas. This must have been an image from his youth when he knew only the burning of coal and all its untidiness.

Operationally, the process of being a scientist in the Griffithsian school was centered on the concept of structure through sampling. Meaningful activity was quickly defined for the advancing recruit as measuring things according to a set of operational rules. Without strict rules for sample taking and preparation, you simply could not know what you were doing. Griff demonstrated the plight that most scientists face after they collect data without using sample design procedures. Because he had seen so many analyses of poorly collected data, he was able to conclude that such analyses only told these scientists that something was wrong. If they would listen, he could tell them how to collect their data next time. Griff preached to us that the material you were sampling had a structure and that it was homogeneous, patchy, or layered. It was our job to find that out and to determine what this structure was correlated with that we could measure. The material we were studying happened to be rocks, which are made up of grains, matrix, and cement.

Operational definitions were provided for each constituent based on a topological nomenclature. The grains had sizes, shapes, sortings, and orientations. All the bulk properties of rocks were related to the mineralogy, the size, the shape, the orientation, and the packing of the constituent grains. These bulk properties included grain and bulk density, porosity, and permeability. It was sample and resample to learn about the fundamental character of the constituents of the rocks. This process was likened to particle physics studies in that you counted and measured constituents. We learned that the grain size axial ratio was nearly a constant, varying only slightly from a value of $b/a = 0.65$. Griff would conclude that all the information on size was contained in one measurement. The concept that the size of a grain was in three dimensions was wrong. Grain size was scalar! The experiments were carried out under the microscope, in the Montoursville gravel pit, on the loess banks at Sunbury, and on the Tuscarora scree slopes. Grain size was grain size. No matter where you found a quartz grain or rock fragment under abrasion, it had the same axial ratio. The shape of grains, their packing, and their orientation all were studied, but not as thoroughly as grain size.

The study of the constituents of rocks took 15 years for Griff and his students to carry to maturity. When I arrived to begin my study, it had drawn to a close. In fall 1962, Griff had set his sights on the relatively new field of mineral-resource assessment. The results of his study of the constituents of rocks, however, would form the basis for most of his lectures for years to come. He took the same atomistic approach that had served him so well in the past with him into his new-found field of research. Oil and gas fields now replaced the quartz grains and rock fragments of yesteryear. He told me that the same problems were there to be solved as before; they just went by different names.

Oil and gas fields have sizes, shapes, orientations, spatial distributions, and values. That was where I would begin. Within a year, we had enough data to determine that the size distribution of oil and gas fields was not easy to characterize. The argument over which theoretical frequency distribution best describes the sizes and contained values of oil and gas fields that we started to wrestle with more than 25 years ago still goes on today. In 1965, Griff concluded that the size distribution of oil and gas fields was a mixture of several distributions. This is still a reasonable conclusion, although we now know that the interaction of discovery process and underlying size distribution is responsible for the apparent mixture. Perhaps our most important result was that the arithmetic mean of the observed field size distribution was not in any sense a typical size. It almost always fell in a lonely position in the size distribution hundreds of millions of barrels away from the closest field.

Griff soon moved on from the study of the attributes of oil and gas fields to the topic of optimization of search strategies for their discovery. It was at this point that I gained my introduction to grid-drilling strategies and entered the world of forecasting in earnest. From then on, my time was spent collecting field size data and determining optimal grid spacings. While I was computing target detection probabilities, Griff was focusing in on the specific objective of conveying the message that grid drilling the entire United States was a useful idea.

The method he chose was to use rhetoric that would provoke the collective conscience of all those involved in the exploration for natural resources who made policy decisions. He was convinced that our methods of exploring for oil and gas fields and mineral deposits were very inefficient. Grid drilling was seen as a technique that could be used to search for all types of mineral deposits at once. Why do we have oil companies exploring for oil and copper companies exploring for copper and zinc companies exploring for zinc? Why not have one exploration effort aimed at looking for all commodities simultaneously and doing it efficiently by using a global strategy? The idea driving Griff to make this assertion was drawn from his study of information theory. He saw drilling on a grid as an equal information network that would extract the maximum entropy (gain the most information) from exploration of the Earth's crust. The objectivity of the grid was likened to the objectivity of a sampling scheme that would produce a random sample.

Over the years, the pros and cons of the Griffiths proposal for the grid drilling of the United States on 20-mile centers would be debated by industrial, academic, and government geologists (Griffiths, 1966, 1967). In the bastion of geology called the USGS, the idea would be disparaged but not totally discarded. Who was this professor from Penn State? Why was he trying to change the way we did geology? It was reckoned that he wanted to take the geologists out of the field and put them to work logging a million miles of core. The then-chief geologist of the USGS, Richard P. Sheldon, was sufficiently charmed by the man and his ideas to put the matter to rest by hiring four of his students.

In general, the reaction to the Griffiths grid-drilling proposal was strongly negative (Herfindahl, 1969). This proposal carried with it more than a threat to the day-to-day manner in which we explored for natural resources. At the center of the negative reaction was the well-placed fear that the proposal, if implemented, would deny mankind a basic source of human dignity. This was the right of an individual to dream dreams, to take risks, and, if successful, to discover El Dorado. This argument for efficiency was perceived as being specious and having been concocted to deny the economic agents the right to prosecute the fundamentals of American free enterprise.

In the years immediately after I left Penn State, I was often asked to what my allegiance was anchored. Was it to the grid or to human dignity? On one occasion at a professional meeting, an oil company geologist became so irate with me that he could not eat his lunch after I had defended some of the merits of the grid. He tried to attack me from the position that drilling on a systematic grid was ridiculous because the required land position could never be put together. I really put my foot in it when I noted that perhaps the right of eminent domain could be used to construct the land positions required to get the job done. I gradually learned to withdraw from these discussions early in their development, but not soon enough to avoid a dozen or so of these fracases. As I look back over these years, I am grateful for the experience gained from these interchanges. They provided good preparation for a career in the forecasting business. It is a rather tame affair today to prepare, present, and defend a forecast of the future rates of discovery of oil and gas or of a metal endowment in comparison with the conflicts I encountered over the Penn State proposal to grid drill the United States on 20-mile centers.

One of Griff's images sticks in my mind. He would often say that you should watch the reaction of your audience when you speak. If they clap politely and give you a nice compliment, you have done absolutely nothing. If they hoot and howl, you are on your way to the truth. If they crucify you, you know that you have found it.

2

The Oil Company Research Laboratory

After leaving graduate school, I was employed by a government contractor, the Geotech Company in Alexandria, Virginia, where I analyzed oceanographic data. When I joined the company in 1967, the staff directory was reprinted, and we were 127 persons strong. Photoreduction was required to make the list of names fit on a single page. After about 6 months, however, I became concerned as I watched this work force being reduced. Friday seemed to be the day people were let go. On the Monday morning after each of these Fridays, I noticed that the staff directory did not have to be reduced as much. When the staff had dwindled to 61, I resigned and went to work for an oil company. I left believing that just about nobody was going to have an entire career and retire normally from the government-contracting business—not from this company and probably not from any contract research company on the beltway that surrounds the nation's capital. My primary goal was to become a research geologist with the U.S. Geological Survey in Washington, D.C., but tight budgets during the late 1960s and early 1970s made that prospect only a dream.

In August 1969, I joined the research laboratory staff of Cities Service Oil Company in Tulsa, Oklahoma. At this research laboratory, the office furniture was not much improved over that of the government-contracting business and its gray steel desks and chairs. The exception was the director's office. He sat behind a large wooden desk that had a thick glass top in a room with paneled walls. This was the visible beginning of the corporation's chain of command and tangible evidence of the corporate philosophy. The style of the work was what struck me the most. Here, your opinion was taken as being equal to an analysis. You did not need to justify how you had arrived at a conclusion. In fact, no justification was wanted. You were hired because you were an expert in a scientific or technical field. The value of your contribution would be determined by how useful it was over time. From the manager's point of view, the daily routine was working best when you contributed your opinion to a problem that had just come in as a request from a field office. I had been with Cities Service only two or three weeks when my boss called me in and gave me my first assignment. It was a problem in forecasting.

I was told that I was being sent to the Rocky Mountain regional office in Denver, Colorado, to discuss a potential application of statistics and computers to an operational problem. I would report to the regional geologist at 7:00 A.M. and was not to be late because it was important to make a good impression. My plane reservations had been made, and my ticket would be brought around later in the day. I asked to be told more details about the problem. The boss said that he had talked to the regional geologist and had told him that I was just the man they needed to solve this sort of problem.

To be sure, my initial impression was good—I arrived early. The cheerful receptionist told me I was expected but would have to wait a few minutes because the regional geologist was busy with the assistant chief geologist from the Apache Drilling Fund Company. The time was 6:55 A.M. The receptionist kept me busy for quite a while by talking about the weather, the plane ride, if I'd ever been to Denver, where I grew up, and so forth. The door finally opened and out came a man carrying an armful of map tubes. He looked as though he had suffered a rather bad setback. To my benefit, I would meet this fellow later in the day.

The regional geologist came out of his office and welcomed me with a big handshake and one of those mighty-good-morning greetings that southerners like to use. I was a New Englander in mind and character, and I did not know what to make of this sort of greeting. His office environment had been carefully prepared to deal with all types of folks, and I would soon learn that, in particular, it was designed to deal with such individuals as the disappointed fellow who had just left. The regional geologist's desk was positioned to dominate the room. In front of the desk and close to the door was a large wooden table. When a visitor entered, he was offered a chair at the table, and the regional geologist sat opposite. The furniture was positioned to the advantage of the regional geologist. As he sat down, he had at his back the expanse of his large room. His visitors had to assume a lesser position, sitting in front of a rack of map tubes, which was against the wall and close to the door.

The regional geologist ushered me to the chair his previous prey had just vacated. Propping his very handsome cowboy boots on the massive wooden table, he asked me where I was from. Suddenly he came to the point: "The manager of the geology section of the research lab says you know all about math and computers. We need some of that out here." The problem he faced was of a making that he really did not like. He spoke in negative tones about a vice-president in the front office in New York who had recently taken an intensive eight-week training course in new methods to manage a corporation at one of those fancy eastern schools. The course covered computer modeling, game theory, market forecasting using econometric models, and present net worth calculations. As he talked, he projected the image that folks in the regional offices, such as himself, were temporarily at a disadvantage in terms of the new vocabulary that was being used against men in his position in the company. He understood that the situation he was in was a war of words. The real issue was that decisions in drilling wildcat oil and gas wells should be analyzed in a modern management science structure. He spoke of the situation as MBA-madness running loose in our corporate headquarters, about which something had to be done.

The regional geologist asked me what I knew about these new words. He watched my response as he went over some of the new vocabulary with which he needed to become comfortable. After a few minutes of discussion, he seemed satisfied. He then said to me, in his mighty-good-morning manner, "Now, here is the deal. We need one of your flashy research lab reports, about so thick." He held up his fingers about one-half inch apart. "And here is what I want your report to say. I want it to say that it is not so bad to drill 121 dry holes in a row in the Rocky Mountains, mostly in the Powder River basin, like we just did. Oh, and, when we make the presentation to the executive vice-president in Tulsa, I do not want any red slides. He hates red slides. If you use one red slide, you will blow the whole deal." This executive vice-president was in charge of all the major decisions made in exploration and production. If convinced, he had the power to stop the crusading vice-president in the corporate headquarters in New York. He had to be *thoroughly* convinced because it was risky to argue against an idea proposed by the headquarters staff.

We left his office and went to a map room where a roll of maps of the study area had been prepared for me. I was next taken from office to office and introduced to the geologists who had been assigned to work with me, and I wrote down phone numbers and memorized the faces to go with them. I suspect that the real purpose of these introductions was to let the staff see who was going to help them in their struggle against the vice-president in New York. After returning to the regional geologist's office, he asked me point blank, "How are you going to solve the problem?" I said I would use a Monte Carlo simulation model. He looked pleased as he repeated the words half audibly, "Monte Carlo simulation model. That will do just fine." I left promising a progress report in a week. He said he would report back to my boss later in the day on how things went.

After boarding the plane for Tulsa that same afternoon, I found that I had been assigned the seat next to the man from the drilling fund who had preceded me in the regional geologists's office that morning. We recognized each other and began to talk about how his day had gone. It had been an absolute failure.

No deal had been turned. He told me that I worked for a really good company but that we drove too hard a bargain. He said that he needed more than the dollar-a-foot dry hole money and sixteenth override that he had been offered this morning at the big wooden table.

I told him about my new project. He seemed hardly to be listening when suddenly he turned his whole body around in his seat. He said that what I ought to do was figure out a gimmick to convince Cities Service Oil Company to buy one section in the middle of each undrilled township in each of the basins in the Rocky Mountain region and then drill just one wildcat in the middle of each of these sections. If I could do that, then the company would find lots of oil and gas, and I would make a real name for myself. He said that the computer could help me do just that. If my man wanted a strategy, I should give him a real one by providing reasons to drill lots of wildcat wells. It sounded as though he was suggesting a grid-drilling strategy with the idea of one wildcat well per township.

Back at the research lab, I put together a random-walk simulation model. The data used for the analysis were gathered by digitizing a map of the oil fields discovered to date in the Powder River basin (Fig. 2.1). A 40-acre spacing was used. I gave little progress reports to our morning coffee klatch, which was attended by most of the research geologists and geophysicists. Pointed ques-

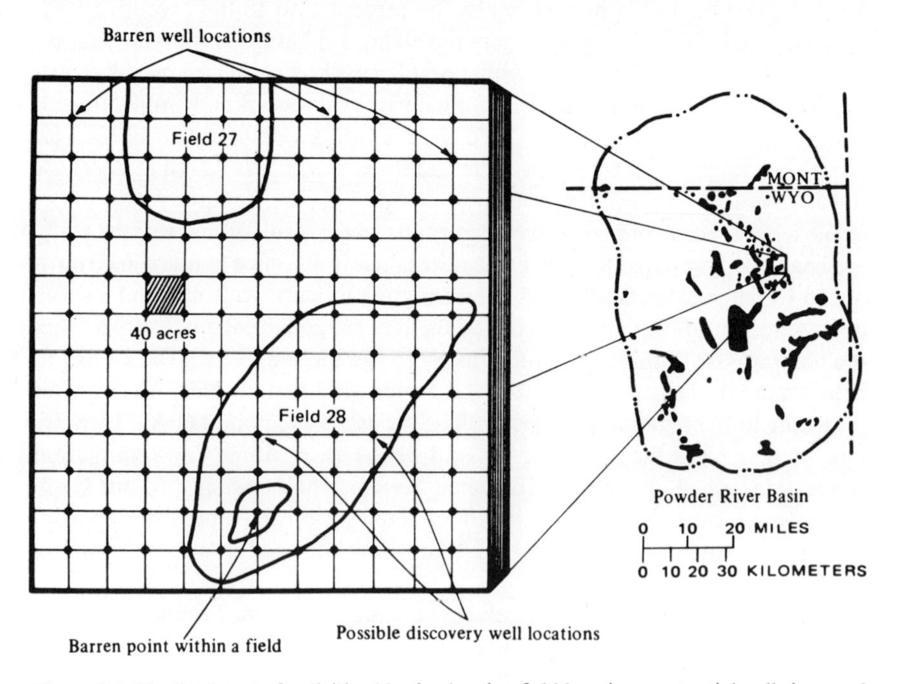

Figure 2.1. Idealized part of a digitized basin showing field locations, potential well sites, and barren points within a field (Drew, 1974).

tions were asked regarding why such a project was being done in the research department. The consensus seemed to be that this kind of stuff was acceptable for the exploration and production planning department to do but was not wholesome for a scientist in the research laboratory. Concern died down somewhat when it was learned that the study had been called for by a field officer.

The random-walk allocation rule was chosen because I knew it would have been self-defeating to base any part of the analysis on my beloved grid drilling. It had become impossible to have a civil conversation on the topic of locating exploratory holes by using any geometric pattern. It did not matter that elementary analyses could demonstrate the merits of exploring by using grids. It was bad manners even to mention the idea. I thought it was silly to ignore the fact that exploration using grid-drilling strategies was, in the aggregate, vastly superior to the patchwork strategy presently being used by the oil industry. However, I had acquired the sense that a massive conflict existed between global optimal strategies and the strategy that individual economic agents were able to use. When I recognized that this conflict was a facet of the conflict between capitalism and its independent economic agents versus socialism and its central planning, I did not mention grid-drilling strategies again but put my square, triangle, and rhombic grids on the shelf to collect dust. I did stop long enough to remember that Professor Griffiths was a fan of central planning for governmental programs and wondered if he had been drawn to the elegance of exploring for natural resources by using geometric patterns of drill holes from a deeper philosophical motive.

I did not dwell on such concerns for very long because I could see that a random-drilling strategy mimicked the way exploratory drilling went in a basin over time. The task I had before me was to compute the probability of drilling 121 consecutive dry wildcats in a row in a basin in the Rocky Mountain region. It was obvious from only a cursory inspection of the map for the Powder River basin that the probability of drilling 121 dry wildcats in a row at random was going to be slight. The three or four largest fields were too large to be missed by this many wells. How was I to tell this to the regional geologist? I could have insulted the man by saying that the results from my computer model showed that a random-drilling strategy outperformed their way of going at it. There was nothing else to do but to play it straight, to perform the analysis, and to present the results.

The following criteria were used to evaluate the random strategy: (1) quantity of petroleum discovered, (2) number of dry holes before the first discovery, and (3) number of discoveries. Frequency distributions for each of these criteria were constructed for a range of drilling program sizes. Initially, the distribution of the number of dry holes drilled before the first discovery was the most important result. The chance of drilling 20 dry holes in a row at random before discovering one of the 154 fields that had been discovered in the Powder River basin by 1969 was calculated to be about 0.75; that is, about three-quarters of the time, a 20-well random-drilling program would yield no discoveries if these

154 fields contained all the oil and gas ever to be discovered in the basin. A 50-well drilling program would have had about a 50-percent chance of making a discovery, and a 121-well program would have had an 83-percent chance. Thus, a random-drilling program would have outperformed the company's record more than three out of four times, assuming all the fields were undiscovered at the start of the simulation. This estimate was almost certainly too low because there were 141 untested townships in the basin at the time I digitized the location of the 154 previously discovered fields. I had no idea, nor did anyone else, with maybe the exception of the assistant chief geologist for the Apache Drilling Fund Company, that a field as big and as productive as Hartzog Draw remained to be discovered under this untested ground. It was discovered in the middle of the basin seven years later in 1976. Many additional fields also have been discovered in the Powder River basin since 1969.

I was faced with a fundamental problem in forecasting. My model was a hindsight simulation model. The meaning and usefulness of the probabilities that it calculated were based on the credibility of the digitized field data. These data were incomplete, and the probabilities of discovery were too small because the model was given the entire basin to explore, and there were lots of fields left to be discovered *in* the basin that this type of analysis assumed were *not* in the basin.

If I restricted the model to only the townships that had been explored, I would still have a variation on this same problem because some townships were intensely drilled, and all the fields occurring had been discovered. Others had only been sparsely drilled. I spent some time wondering how I was going to talk to the regional geologist about this topic while maintaining a positive attitude about his problem. How conservative would the probabilities be that I was going to compute? I suspected they might be underestimated by a factor of 2. This estimation problem and many allied problems would form an entire field of research beginning in the mid-1970s. The most commonly used designation for this field of research is petroleum-resource assessment. A subfield of discovery process modeling grew up around the very problem with which I was wrestling as I tried to compute probabilities of discovery with a hindsight simulation model.

By late fall 1969, I had prepared a research laboratory report and had been mindful of the bulk that had been asked for by the regional geologist three months earlier. After it had been printed, advance copies were sent to Denver. A time was set aside for my presentation. Slides were prepared of all the figures and tables, with special care to avoid any trace of red. The presentation went off without a hitch. A stack of copies was handed out, and everybody was polite. After the presentation, I was taken to see the senior geologist. He said that he enjoyed my talk and reading my report. He then spent a long time talking about the importance of making and using oil show maps. Could I help him make oil show maps? I said I could but was confused because his request was not connected to the report I had just presented. What did making maps of shows of oil and gas in abandoned drill holes have to do with drilling strategies? The time came to leave, and he took me to the elevator. The door to the regional geologist's office was closed as we walked past. The receptionist bade me a cheery goodbye. At the elevator, the senior geologist wished me well and said he hoped to see me again soon.

The first thing the next morning, I was questioned very specifically by my boss about who was at the meeting, who asked questions, and what kinds of questions were asked. He took notes but did not say much in response to my answers. After this interrogation, I asked a colleague what to make of it all. He listened to the entire transaction and said only that working for an oil company was sometimes like being faced by a big ball of aluminum foil. You make your impact by pounding a dent in one side of the ball, but, after you turn away for a minute and then turn back to admire what impact you have had, you notice that someone inside the ball has pounded out your dent. I guessed then that I was to understand that my report had been swallowed up by the organization, and I was wished better luck next time.

The next month was spent in planning a long-term research project on quantifying stratigraphic correlations. The project was to be based on electric log characteristics and guided by paleoenvironmental considerations. I picked up some gossip at a local geologic society meeting that a chap in another oil company had suffered a worse fate in his company than I did when he performed a similar analysis. I was advised to do mainstream research on well-defined geologic problems.

A few weeks later, I made a trip back to the same regional office in Denver to learn well-logging techniques from a log analyst and to collect data for my new research project. The office manager, who was the regional geologist's supervisor, just happened to stick his head out the door as I walked by. After inviting me in, he pushed a package across his conference table at me and asked if I had ever seen a big submittal before; he suspected that I had not. It was a legal document, typed in the form of a deed, full of terms I had never seen before. The submitting company offered a 25-percent working interest in 351,000 acres in the southern portion of the Powder River basin for $6 million. Along with the working interest came 50 wildcat wells drilled to the top of the Dakota Formation, turnkeyed to the casing point. The Muddy Formation was the prime objective. At this junction, the office manager invited the regional geologist in, who then took over the conversation.

The regional geologist said that the deal as offered was too big for us, so we would try analyzing about one-half of it along the main Muddy Formation trend. He turned on his southern folksiness as he asked if I would help them out and analyze this deal for them. He said that they would have the exploration and production department do the economic analysis. I kept a poker face and started to play the hand I was being dealt. They asked what I needed from them. I said that a data base of undiscovered fields had to be assumed into position in the drilling simulator. I peddled the idea of using the standard operations-research device of a high-low-median input to cover the range of uncertainty in the assumption of the data base. To my utter amazement, they were enthusiastic. The last impression I had was a closed door and being put in an elevator and sent on my way. I had to wonder if the economists and the engineers in the planning department were behind this new transaction. This department was aggressively pursuing the application of the new management science techniques to the company's exploration and production activities. Using a geologist from the research laboratory to compute probabilities of discovery for them in their economic modeling would be a way to further their goals.

The two office managers seemed to be having fun as they picked data bases for a worst case, a best case, and a median case (Fig. 2.2). They asked me how long it would take to do the analysis. I said maybe it would take a month or two. One month was fine! They seemed anxious to see what their choices of data base looked like dumped out the other side of a discounted cash flow analysis.

The exploration and production planning staffer with whom I worked on the project had an immediate sense of what we were going to do, and we nearly made the one-month deadline. In the process of preparing the report, we faced the scrutiny of a chemical engineer who had been asked by the director of the research laboratory, who was also an engineer, to evaluate our modeling effort. He found no fault with the economic analysis. He made only one remark during the meeting, which was that, with all our good engineering and economic data and analysis, it was too bad we had to work with such a poor characterization of the undiscovered oil and gas fields in the area under study. The engineer knew how to calculate the risk in an engineering system design and how to minimize the consequence of risk to the company's cash flow. It seemed to him that we geologists were very primitive in our ability to do the same in the exploration business. We explained the status of the *ex ante* geologic knowledge before drilling a wildcat well. He shook his head in disbelief and was sure that it was not too difficult a problem to be solved correctly. However, he never offered to work with us to understand the problem we faced.

The reward for our report was the receipt of another large submittal for analysis, this time from the Burlington Northern Railroad. The only feedback on our first analysis came several years later from an exploration geologist who had worked in the Denver office at the time we had done the work. He said that the company should have taken our advice and bought the deal outright because a lot of oil and gas was found on the acreage in that submittal. I argued with him that it was the regional geologist who had done the most important work because he had assumed a three-case scenario that later proved to cover the outcome of drilling the test area. The exploration geologist did not care to listen to my argument. He was sure that a blunder had been made, and there was no changing his mind on the matter.

As I look back, the analysis of this first submittal marked the beginning of

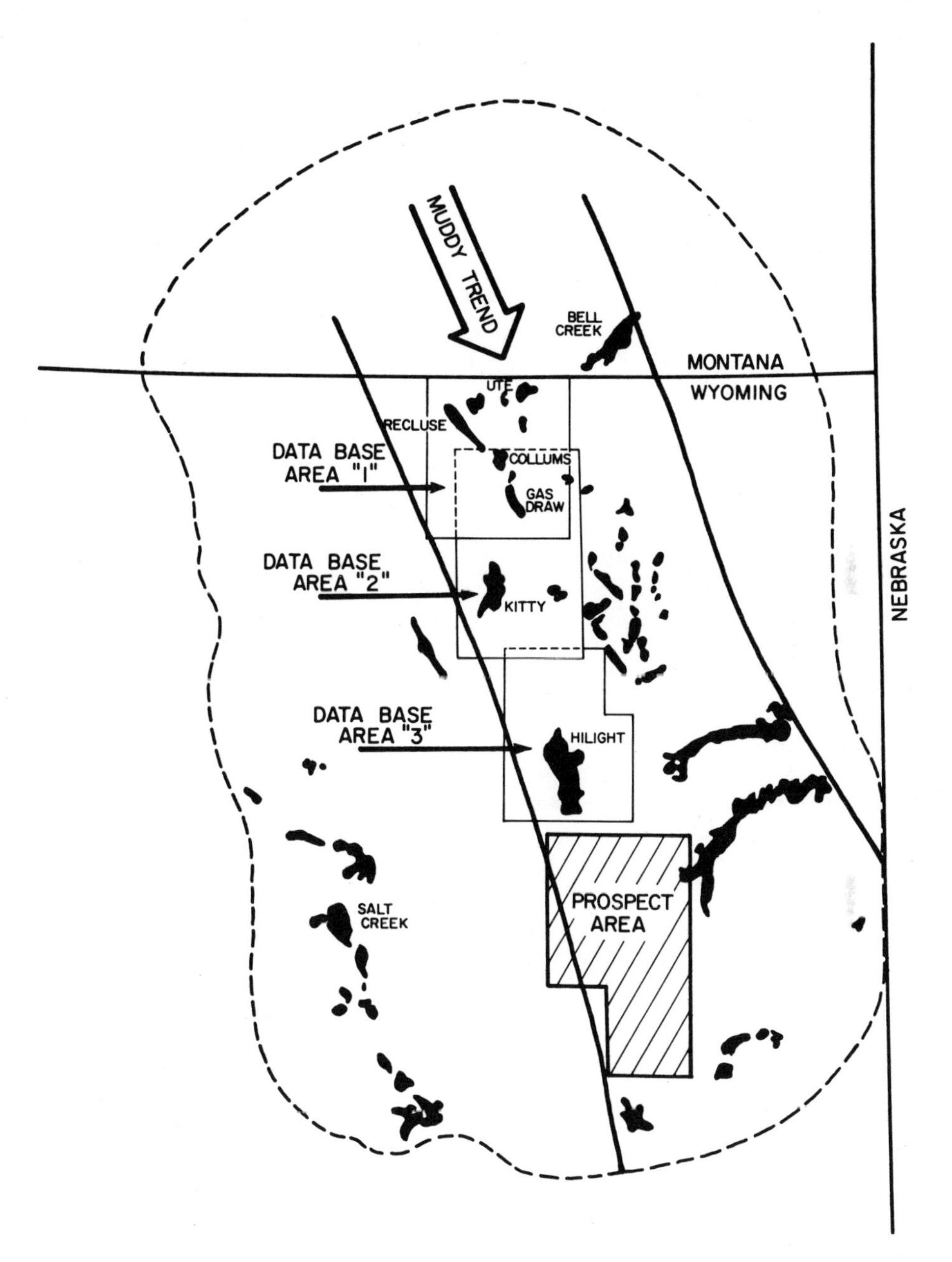

Figure 2.2. Location of prospect area and three data base areas in the Powder River basin (Drew and Campbell, 1970).

the formation of my attitude on how the exploration process ought to be modeled. By its very nature, this submittal had driven the analysis down to the level of the exploration play. The oil and gas fields being sought were located within a single regressive-transgressive cycle of a clastic wedge. French research workers would later coin the term "petroleum zone" for such a collection of oil and gas fields (Bois, 1975; Gess and Bois, 1977). They formally defined the petroleum zone as being any volume of rock-containing fields that have a common source, thermal, transport, and trapping history. In the rest of the world, the term "exploration play" would be more widely accepted. The terms "exploration trend" and "fairway" are often used interchangeably with exploration play, although, to me, they describe slightly different ideas. Exploration trend and fairway seem to me better used to define the real estate in which an exploration play unfolds. The exploration play is a term I would reserve for a collection of temporal events, which include the drilling of a suite of wildcat wells in an exploration fairway and the group of oil and gas fields that these wells discover.

I learned from the analyses of submittals that there are business cycle aspects to the exploration play in addition to its physical characteristics. Within these cycles, the sizes of the fields discovered change in a systematic pattern across the life of the play. The analysis of these patterns of discovery as a function of the number of wildcat wells drilled provides a basis for forecasting future rates of discovery. Many schemes have been proposed since the first formulation presented by Arps and Roberts in 1958. Collectively, these schemes have come to be called discovery process models. Their development and the study of their performance have become the basis of an active research field starting in the early 1970s (Ryan, 1973; Kaufman et al., 1975; Drew, 1975; Barouch and Kaufman, 1977; Drew et al., 1980, 1982; Meisner and Demirmen, 1981; Lee and Wang, 1983; Forman and Hinde, 1985).

My last project at Cities Service Company was in the minerals division. Although nobody said so at the time, this project was an attempt to salvage a philosophy of management. During fall 1971, the managers of the mineral division were being taken to task by the same vice-president in New York who had caused my path and that of the regional geologist in the Denver office to cross two years before.

This vice-president had recently taken another of those very expensive short courses geared for upper management types and had learned that market share was an important concept in the management of large corporations. The minerals division was easy prey because it was diversified into many small activities, which, even under the best economic conditions, would generate small revenues. All around this country and in Canada, there were small offices staffed with as few as a single exploration geologist. The vice-president in New York saw the minerals division as fertile ground for the application of his newly acquired management skills. He did not want to make a frontal assault, so he commissioned a management consulting firm to make a study of the uranium exploration and production projects the mineral division was supervising. He said that he intended this study to help add direction to the far-flung activities of the minerals exploration group.

This study was targeted at the uranium exploration and development activities of the minerals division. The vice-president did not go after the large copper exploration activity nor did he pick out the smaller exploration activities for sulfur or zinc. His goal was to disband the entire minerals activity and to shift the exploration monies to the oil company. He had decided to use a war of attrition, and picking off the uranium exploration and production activity was about as big a chunk as he could manage. He knew the smaller activities would fall victim to a market share analysis if he could use it effectively against the uranium exploration and production activity.

The consulting firm delivered its report, which said that the company's current market share of the uranium was very small and that the cost of expanding it to an acceptable level was very large. This conclusion was no surprise to anyone. It did not matter that the projected bottom-line profit margin on one of the pilot plants was very high. With the help of the consulting firm's report, the vice-president was able to form an effective image of the officers of a major corporation as not being able to maintain or expand their income and profit using small projects, no matter how profitable or how many. The minerals division yielded to the pressure and sold off its uranium activities.

In the next few months, the company went out of the zinc exploration business by closing its office in Jonesboro, Arkansas. The managers of the minerals division recognized their predicament and made the decision to draw the line at exploring for porphyry copper deposits, thereby keeping the Salt Lake City, Tuscon, Ely (Nevada), and Vancouver (British Columbia) offices open. This decision was based on the hope that the cash flow generated by production from a porphyry copper deposit might be large enough to be noticed at the corporate headquarters.

The idea of using an internally generated study to combat those to come from outside consulting firms brought the managers of the minerals division to the oil company research laboratory. I can hear the director of the research laboratory as if it were yesterday making the sales pitch, "We have a combat veteran from our own version of these sorts of battles here in the laboratory." A deal was struck, and, for the next seven months, like a game of tag, I was "It." I was assigned to a project coordinator in New York City and was given a large cash advance and permission to make my own plane reservations, which were charged to an account number. When I asked why I needed so much money, the answer was that, when my phone rang, I would always be ready to go.

It did not take long to see that the minerals business was very different from the oil business. A much larger effort was placed on the evaluation of prospects. The staff members spent much more time analyzing each stage of the process before they drilled a hole. The culture of their business was different, and it was my job to model it. The Monte Carlo simulation model that I used to analyze the activity of exploring for porphyry copper deposits contained a chain of steps. A decision tree type of structure was used to evaluate the effectiveness of different sizes of exploration budgets (Fig. 2.3; Table 2.1 shows the expected performance of a specific budget).

The values of the transition probabilities in this chain had to be estimated by interviewing the geologists in each of the offices. The distributions that described the outcome at the end of each stage were estimated from tabulated historical data and expert judgment (Figs. 2.4–2.8). Several of the geologists I interviewed were not pleased with me or the other members of the modeling team because they did not see any merit in our strategy. The intensity of their feelings became clear to me in Salt Lake City, when one geologist, who had fire in his eyes, restrained himself from bashing me with a pool cue as he vented his feelings against the vice-president in New York and the management of the minerals division for being weak in their defense of how minerals exploration was being carried out by men like himself. Nothing I said that evening could convince him that I was trying to help. After we separated at about 1 A.M., I thought about calling the project coordinator at home in New Jersey (3 A.M. his time) to keep him informed about what was happening on his project in Salt Lake City. I thought better of the idea and saved my remarks for a few days later when I told him that he almost got a call in the middle of the night. He said that he knew it was going to be a rocky road with this geologist and thought that the situation called for small-group dynamics. So he decided to

Table 2.1. Expected Performance of a $10-Million Exploration Budget

Activity	Cost symbol[a]	Number of times activity is completed	Expenditures ($1000)	Percentage of budget
Raw prospects developed	CO	173	692	6
Raw prospects field checked	C1	173	191	2
Land acquisitions	CLR-CLD	86	950	9
Detailed surveys	CS	86	863	8
Submittals retained	SO	170	100	1
Prospects	S3	9	5	<1
Drillable targets	S2	3	16	1
Mineral deposits	S1	0	0	0
Targets drilled	CD	67	3597	34
Validation drilling	CDGT	5	1839	17
Shaft sinking and sampling	CSS	1	2203	21
Total budget			$10,456	

[a]CO is the cost of internally generating a raw process; C1 is the cost of field checking a raw process; CLR is the cost of acquiring land; CLD is the cost of acquiring additional land; CS is the cost of detailed geophysical survey; SO is the cost of examining a submittal; S3, S2, and S1 are field costs of checking submittals; CD is the cost of drilling a target; CDGT is the cost of drilling for grade and tonnage; and CSS is the cost of shaft sinking (or its equivalent) and bulk sampling.

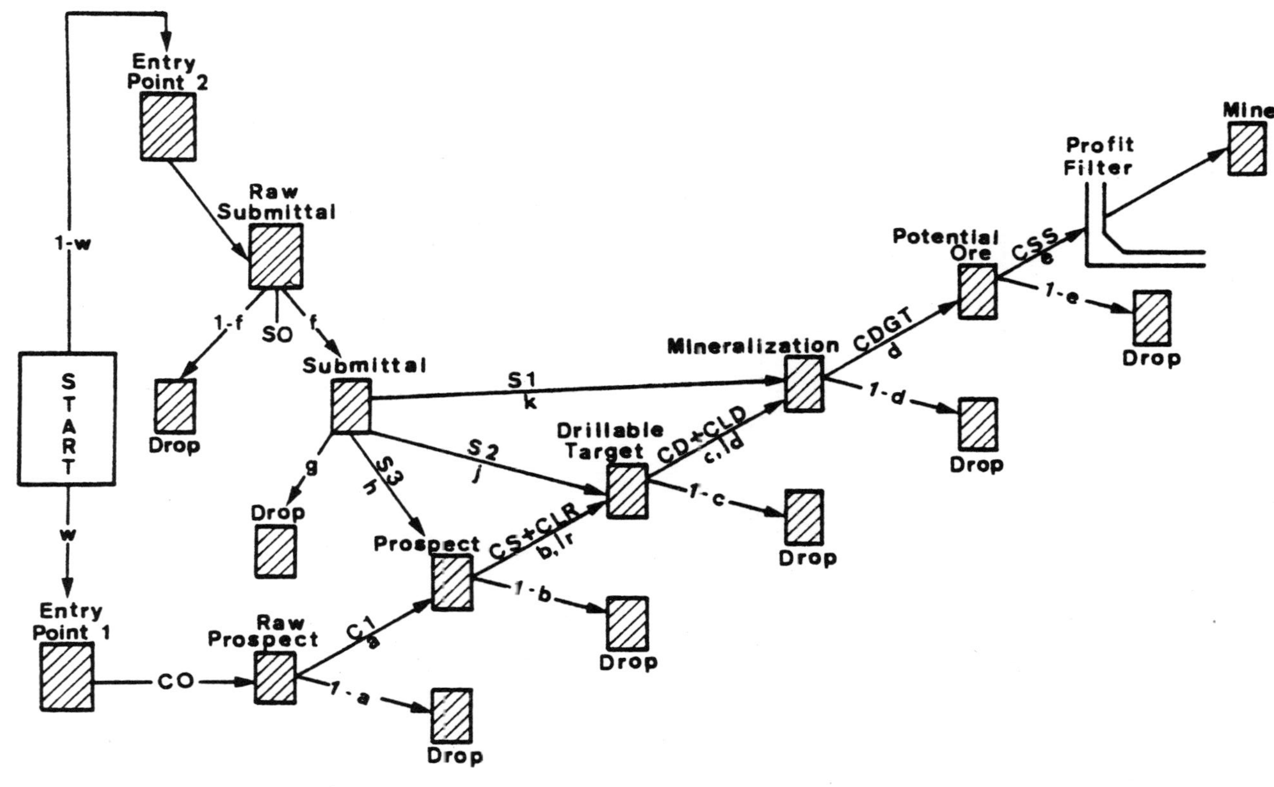

Figure 2.3. Exploration model flow chart (Moore and Drew, 1979).

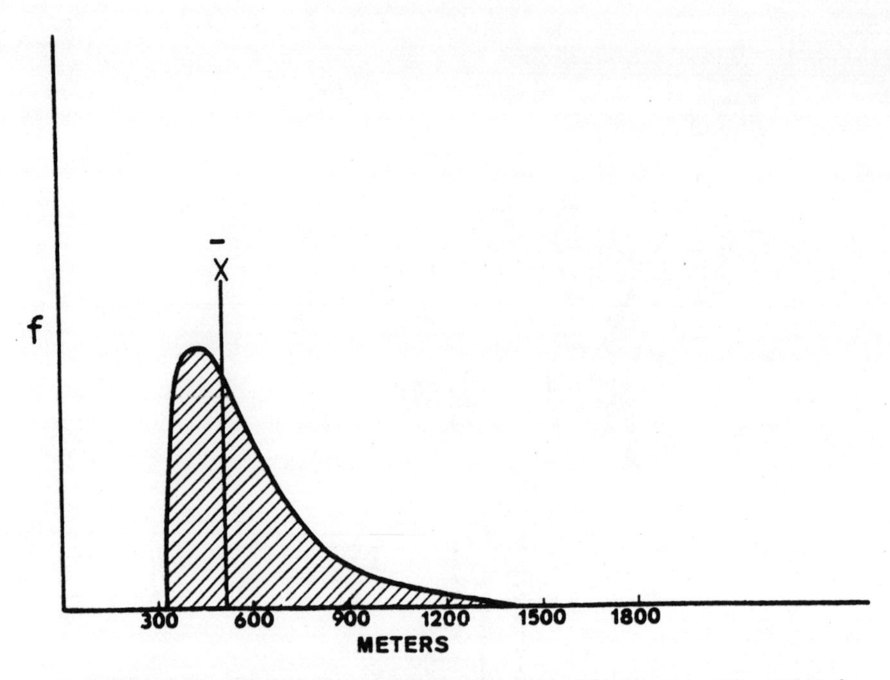

Figure 2.4. Frequency distribution of the average depth to which a target will be drilled. f = frequency and $\bar{X}$ = mean depth.

stay home and leave it to me. I chuckled at his defense as I called him a chicken.

A report on the project came together at the level above me. The economic modeling for the project was done in the oil company's treasury department. It was all hush-hush as I handed over my part of the report. Because the economic analyst from the treasury department felt that I did not need to know, he would not show me the output of any of my simulation runs after he had processed them through his profitability index filter. I was curious about this filter because there was an awful lot of information in it about the economics of exploring for and mining porphyry copper deposits. It would have been acceptable to me if I could have inspected the figures. He could have stood over me and watched, but it was no use even asking because the filter was a secret, and that was that. I did not stew over the treatment of my request to see the final product of my work, which had been integrated into a form for presentation to the top management, because my four-year wait for a job offer from the U.S. Geological Survey ended.

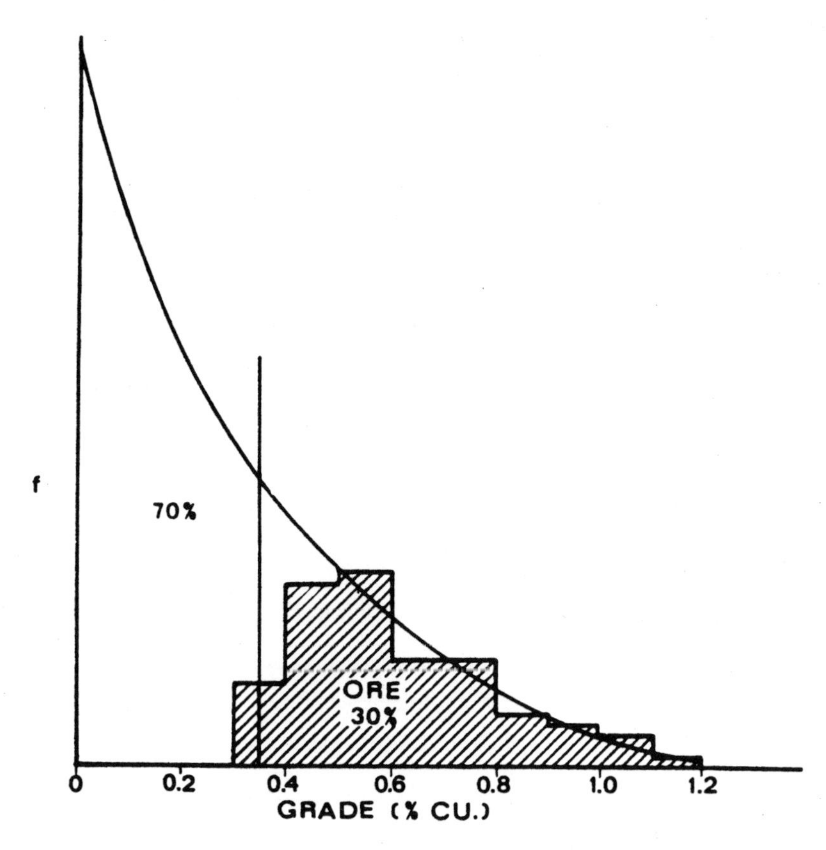

Figure 2.5. Frequency distribution of the grade discovered in mineralized properties.

Figure 2.6. Frequency distribution of ore tonnages.

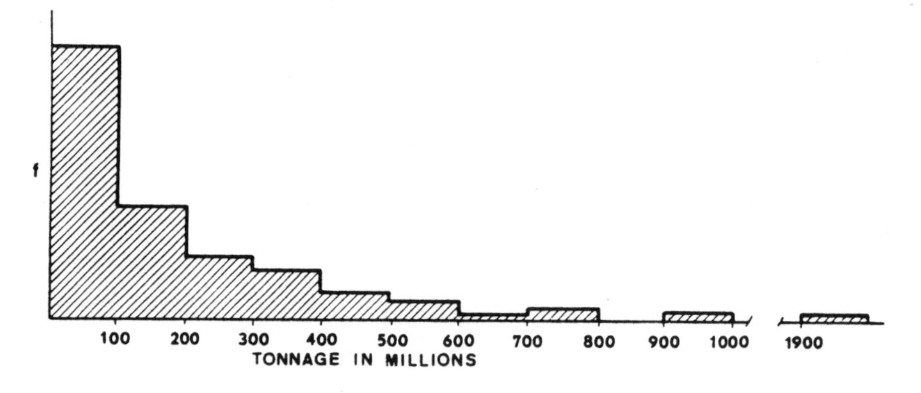

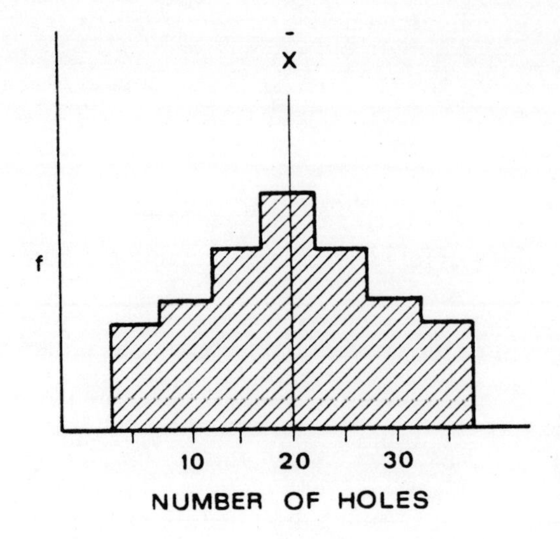

Figure 2.7. Frequency distribution specifying the number of validation holes to be drilled on a property.

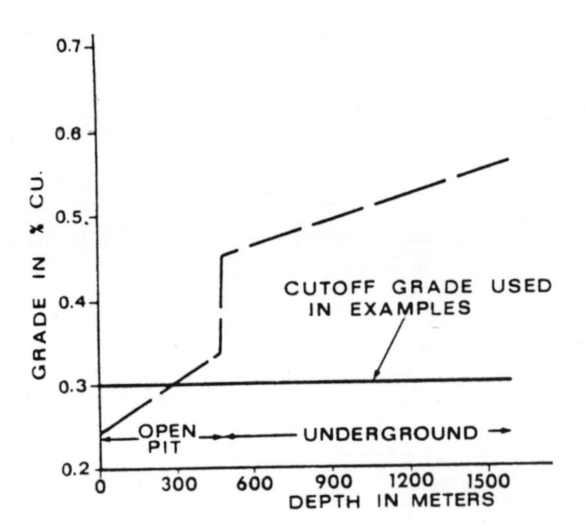

Figure 2.8. Relation between cutoff grade and depth after target drilling.

3

The U.S. Geological Survey

The long-awaited call from the U.S. Geological Survey (USGS) came on June 8, 1972. A government hiring freeze was off, and a new branch was being formed to study assessment methods to estimate the quantity and quality of the nation's undiscovered mineral and mineral-fuel resources. A pleasant part of adapting to the way of life at the USGS in Washington, D.C., was learning that there had been a long history of scientific contribution to the field of resource assessment. In earlier years, these achievements were mainly the work of scientists who had other responsibilities and worked away at resource-assessment techniques and their application as parallel efforts or as special assignments. Some of these individuals were Foster Hewitt, Tom Nolan, Sam Lasky, Tom Hendricks, Al Zapp, Vince McKelvey, and King Hubbert. After more than 35 years of contributions by these and other research workers, the decision had been made to focus attention on this field of research and to house it within a single branch of resource analysis. Although no one realized it then, the timing of this decision in early 1972 was fortuitous because of the oil embargo that occurred in 1973 and the subsequent energy crisis.

When I arrived, Vince McKelvey was director of the USGS and M. King Hubbert, although he would soon retire, was at the zenith of his career. Hubbert was writing and lecturing extensively about his forecast that the ultimate U.S. production of conventional crude oil would be 170 billion barrels. By using a different procedure, McKelvey and his coworkers had arrived at a much larger estimate of 560 billion barrels. These two men with their very different estimates had became well-known antagonists sometime during the late 1960s. Although McKelvey became the director of the survey in 1972, he maintained a strong interest in the field through the mid-1970s and afterward during his retirement.

When we moved into our new headquarters in Reston, Virginia, in fall 1973, I was surprised to find that I was occupying an office only a few doors away from Hubbert's. Although I treasure the opportunity of getting to know Hubbert, I must say that some of the most tiring hours of my life were spent in his office. I would go in to ask a simple question, but once Hubbert had me in his office, he would make me sit beside his desk as he replied in long, convoluted monologues. I spent many hours sitting at mental attention in this position. Sometimes I was surc that only being in traction in a hospital bed could have been worse. It was my curiosity that made me go to his office and ask him how he made his by-then-famous 1962 forecast of ultimate U.S. productivity, arrived at the date of peak production, and made the expanded forecast made in 1967 (Figs. 3.1–3.4). I wanted to know in more detail than he provided in the short paragraph on page 59 of his 1962 paper how he had fit his data. It was not until 1982 that he published the type of account for which I was looking. In that paper, Hubbert revealed how he prepared the 1962 forecast and the later forecasts of 1967 and 1974 (Fig. 3.5; Hubbert, 1982).

In 1974, he seemed reluctant to discuss his methods, and he told me over and over again that he had plotted the data. He said it with such force that I was never quite sure what else he might have been trying to say. I gathered that I was supposed to go away convinced that the last word had been said. Instead of using good manners, I came back again and again to ask him procedural questions on how he had plotted the data. For nearly a year, he took me on one nostalgic trip after another through the nooks and crannies of Shell Oil Company, where he had worked for more than 20 years, discussing the people he had known and his experiences at the Survey. One of Hubbert's favorite

Figure 3.1. Estimate of ultimate U.S. production of crude oil (Hubbert, 1962).

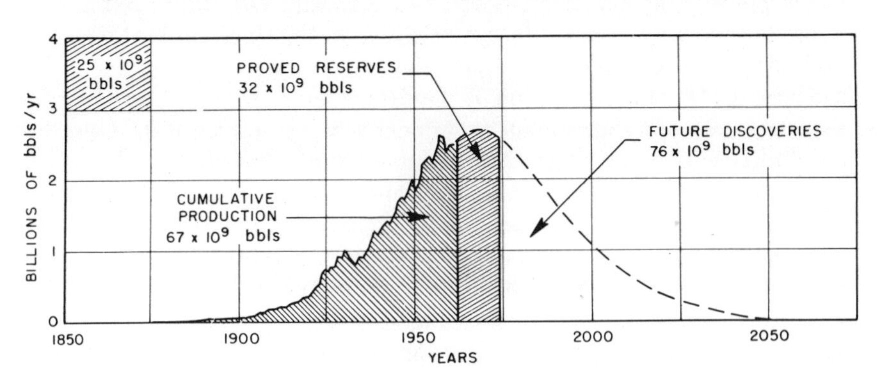

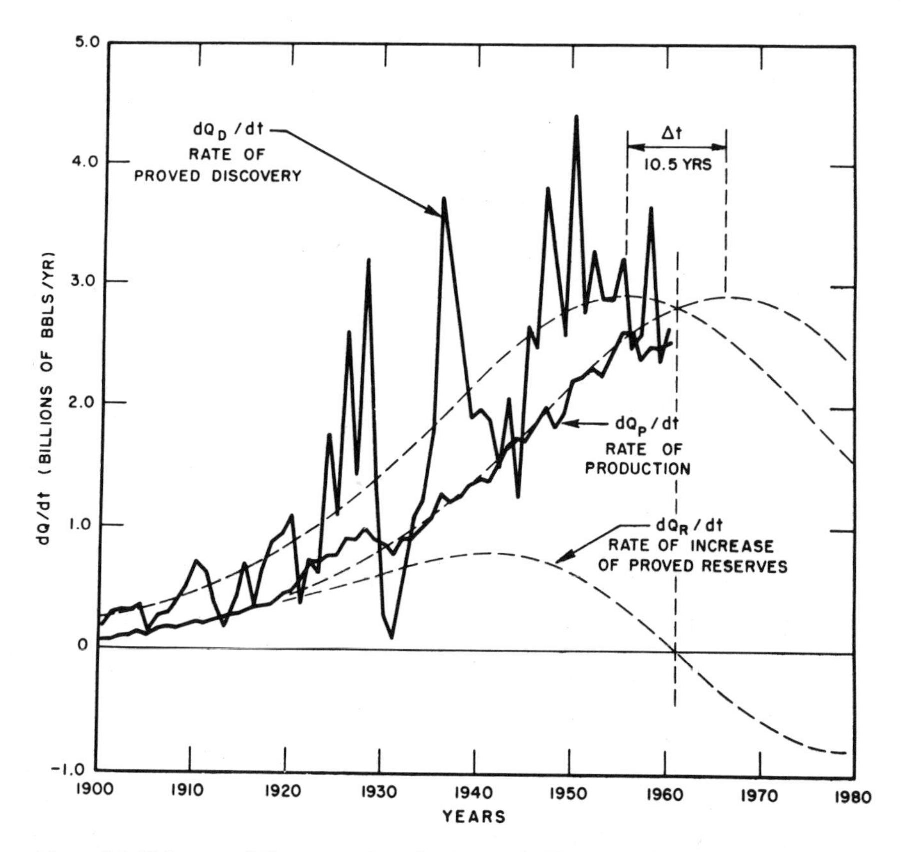

Figure 3.2. U.S. rates of discovery, of production, and of increase of proved reserves. Dashed curves from analytical derivatives (Hubbert, 1967).

stories was one he told about an incident that occurred at a professional meeting. He walked up to a group of people who were discussing whether or not he was still alive. Even after introducing himself, one member of the group looked at him squarely and averred that he was quite certain that Hubbert was dead. Hubbert told this anecdote with great pleasure.

During our long conversations, he would frequently talk about the idea that scientists had to develop clear arguments. His used the legal profession as an example of how not to present scientific conclusions. It was irksome for Hubbert that lawyers are allowed to win cases based on the axiom that the best argument wins. He believed that a scientist seeks the truth, whereas a lawyer just wants to win. Hubbert believed that being justified in a legal sense meant nothing. To him, it was a bad business to use a legal proof because you could find yourself voting on what 2 + 2 is supposed to equal and the jury could very well bring in the verdict that the answer is not 4. He used this legal metaphor

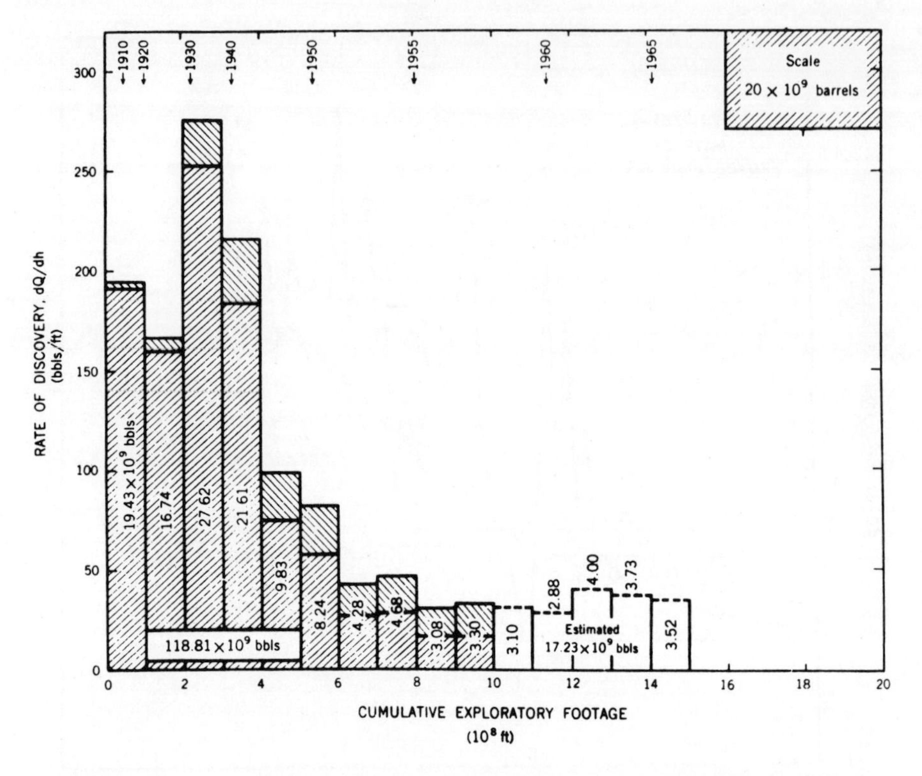

Figure 3.3. U.S. crude oil discoveries per foot of exploratory hole, averaged for each 10^8 feet, versus cumulative exploratory drilling. For first 10 columns, lower shaded area represents the National Petroleum Council (1965) estimate as of January 1, 1964; upper shaded area represents oil added by correction factor α. Last five columns are based on annual American Petroleum Institute estimates of oil added by new discoveries and increased by a factor of 5.8 (Hubbert, 1967).

frequently as he talked about his battles with the men who ran the USGS. Of all the battles, there was one in which he believed most deeply. At the time, the official position of the Survey was that the ultimate U.S. productivity of crude oil would be between 500 billion and 600 billion barrels. Consequently, Hubbert's estimate of 170 billion barrels was too far below this range to be reconciled in any reasonable manner.

Each time I talked to him, my question was the same, although I only asked it outright three or four times in the course of more than 20 trips to his office. I wanted an explanation of the procedure he used to obtain his forecasts—exactly what data he started with, how he proceeded, and, in particular, the mechanics of how Q_∞, the ultimate productivity, was chosen. He knew what I was after, and I could not understand his reluctance to show me. In place of

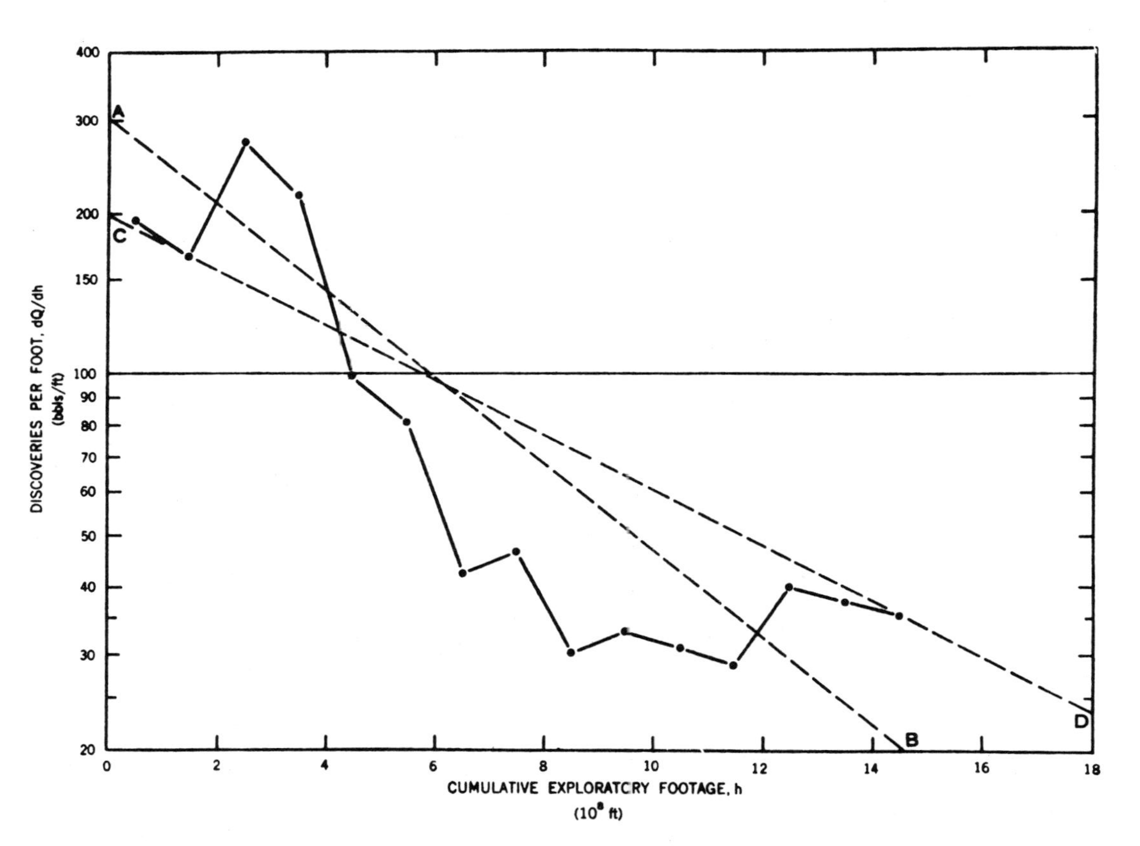

Figure 3.4. Data from Fig. 3.3 plotted on semilogarithmic paper (Hubbert, 1967).

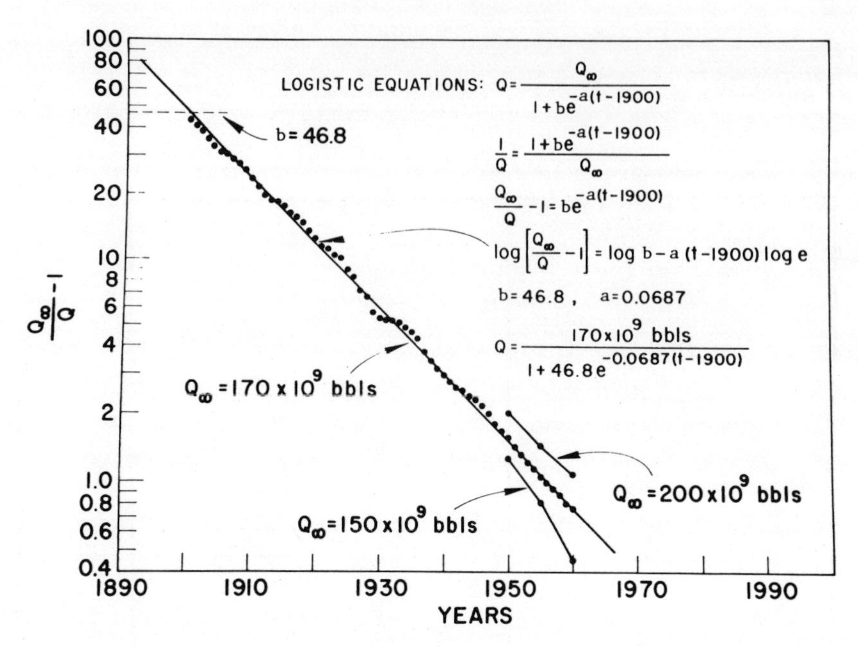

Figure 3.5. Graph method used in 1962 for determining the constants of the logistic equation for U.S. cumulative crude oil discoveries (Hubbert, 1982).

answers to my questions, I acquired a massive amount of information about his life and times.

Then, one afternoon, for no apparent reason, he called me in and took me to his map case. He pulled open the top drawer and took out a pile of paper. There were tables of data and plots on graph paper and also a piece of mylar that had several parallel ruled lines on it. He took one of the graphs and laid the piece of mylar over it and showed me how he chose a Q_∞ and solved for the line. Hubbert's method consisted of choosing a value and solving for points. If the resulting line made up of these points curved away in one direction, he knew his choice was too high, and, if it went the other way, it was too low. He repeated this procedure until he obtained a straight line. I bit my tongue to avoid saying anything that would imply that I thought the procedure was simple.

The history of science is full of important breakthroughs made the same way Hubbert's was by using only a piece of graph paper and having an idea that nobody else had ever tried. After the stroke of intuition, there are always those who swarm in to criticize and/or refine the new idea. David Root, who was Hubbert's assistant at the time, sized up the importance of his contribution when he said, "To appreciate Hubbert's contribution, you must look at when he did his work. He had it all figured out in the 1950s when he saw the first derivative change in the production curve. Just remember that date because that is what is important about Hubbert."

There is a curious fact about Hubbert's work that is not widely known: he never intended to make a forecast of the ultimate productivity of crude oil in the United States. He meant only to forecast the date of peak production. His decision to emphasize the forecasting of ultimate productivity came in the late 1960s when he connected the results from his early curve fitting with those of the extrapolation of the discovery decline rate curves. It was the linking together of the results from these different concepts that created the potential for criticism. Hubbert asserted that his ultimate productivity estimate of 170 billion barrels (Fig. 3.6; Hubbert, 1974, p. 126) had been thoroughly justified because he obtained the same number from an extrapolation of discovery rates, which he claimed was an independent method of analysis (Fig. 3.6). He gave sole credit to Al Zapp for developing the basic calculus for the decline extrapolation method (Hubbert, 1967, p. 2215; 1974, p. 101). When Hubbert fitted an exponential decline curve to the U.S. discovery rate data, he obtained an estimate of 172 billion barrels by integratng the area under the curve (Hubbert, 1974, p. 125). He considered this result to be proof that his previous conclusion was correct. In the 1967 paper, he had offered a similar argument in which he used the data presented in Fig. 3.2 to estimate the ultimate recovery of 170 billion barrels of crude oil and the data represented in Fig. 3.4 to estimate a range of 153 billion to 163 billion barrels. I believe that most readers took this conclusion to be a somewhat weaker statement than that in the 1974 paper.

Although I have never had much of an urge to criticize Hubbert's methods or conclusions, I see that an error was made when he claimed that his original conclusion was supported by the close agreement of the 170-billion-barrel estimate versus the 172-billion-barrel estimate produced by two very different forecasting methods (Figs. 3.1 and 3.7; Hubbert, 1974). The problem with this claim, which is obvious to the statistical analyst, is that the discovery rate data did not fit his assumed exponential decline model very well, if at all. A simple regression residual plot would have shown that the model was misspecified. Inspection of Fig. 3.7 reveals the massive lack of fit. It was this lack of fit that DeVerle Harris, professor of mineral economics, University of Arizona, noticed, criticized, and corrected (Harris, 1977, 1984). Harris refit the data and demonstrated the magnitude of the sequential bias introduced by Hubbert's procedure (Fig. 3.8). The point was made that, as time passed, this method would proceed to produce larger and larger estimates of the ultimate productivity of the United States. Therefore, the agreement between the results of this method and those from the earlier method Hubbert had used was purely fortuitous. He also pointed out that, depending on when you made the forecast, the ultimate productivity could be less than past production plus known reserves (Harris, 1984, p. 36). This analysis was well done, and even David Root had to admit that Harris's analysis was valid. This conclusion was given

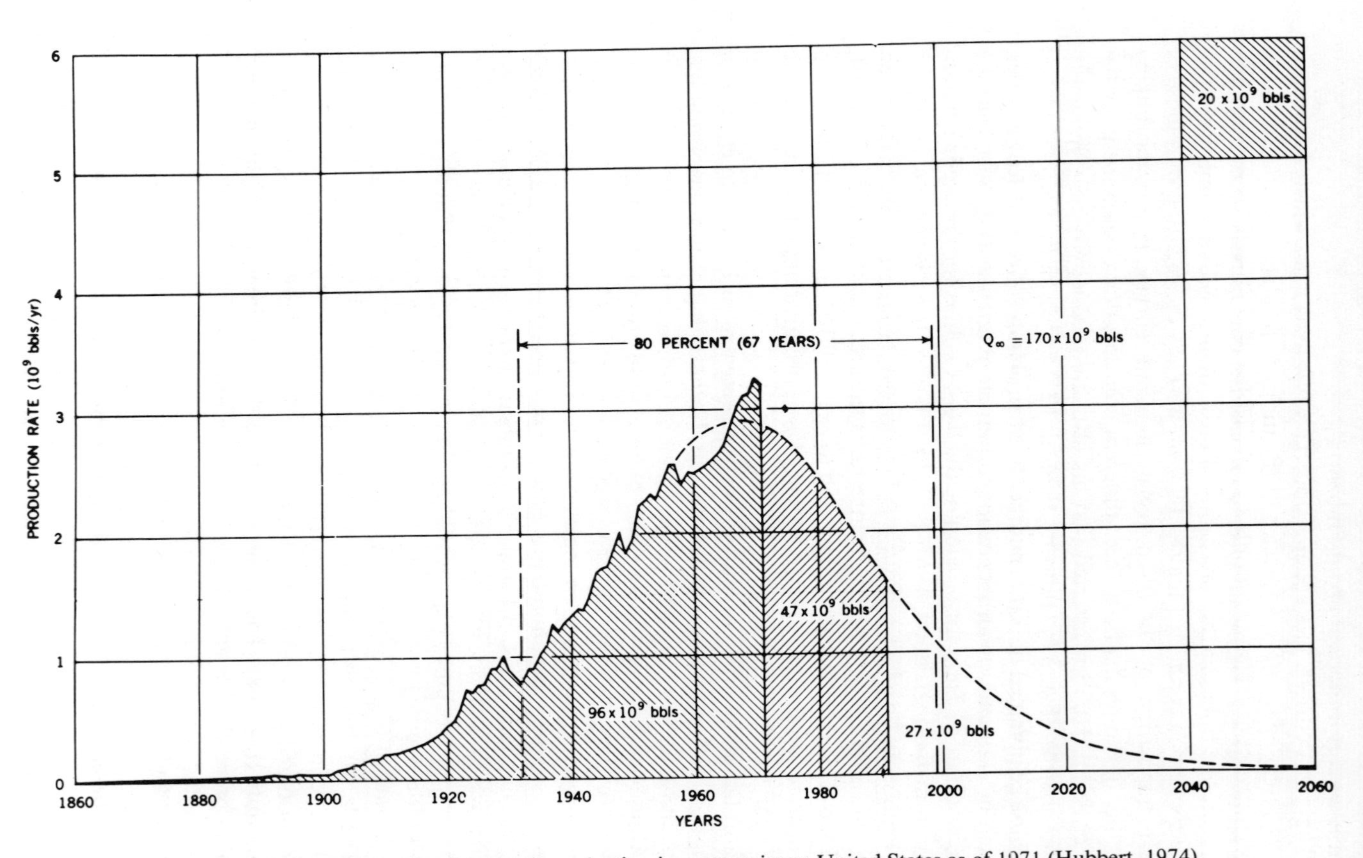

Figure 3.6. Complete cycle of crude oil production in conterminous United States as of 1971 (Hubbert, 1974).

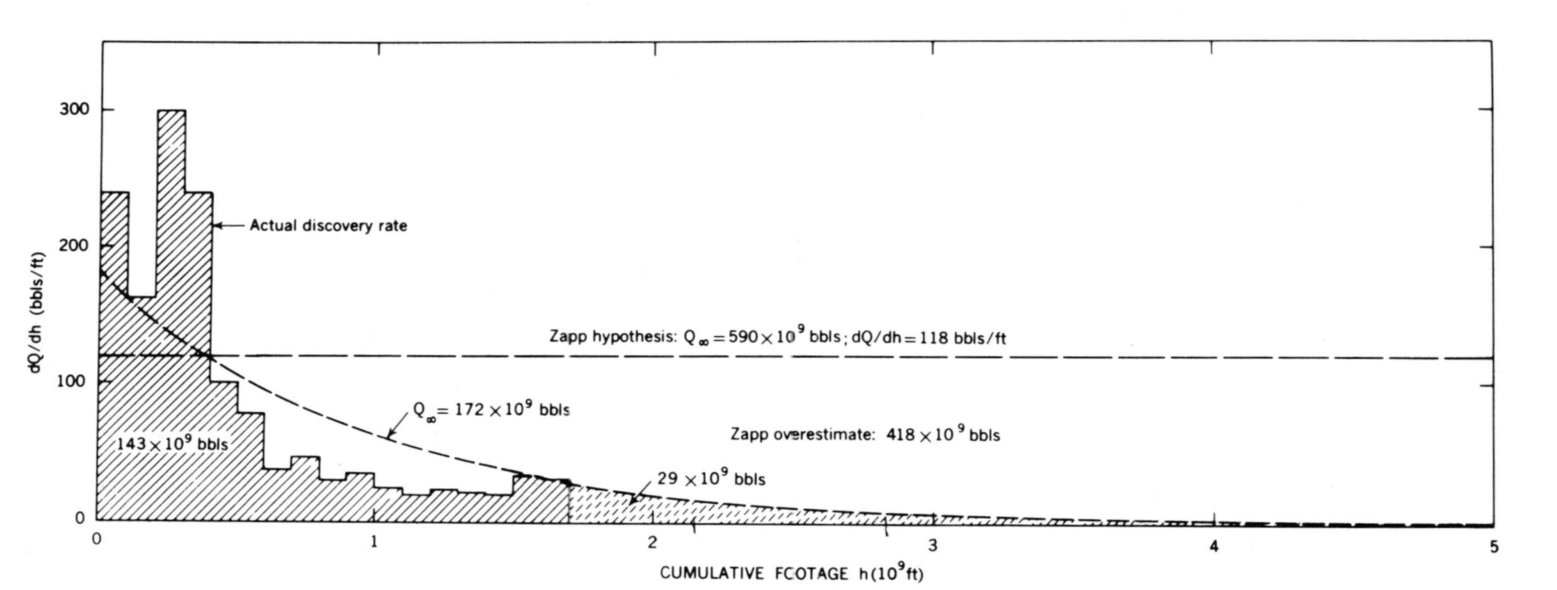

Figure 3.7. Estimation of ultimate crude oil production of conterminous United States by means of curve of discoveries per foot versus cumulative footage of exploratory drilling, and comparison with Zapp hypothesis (Hubbert, 1974).

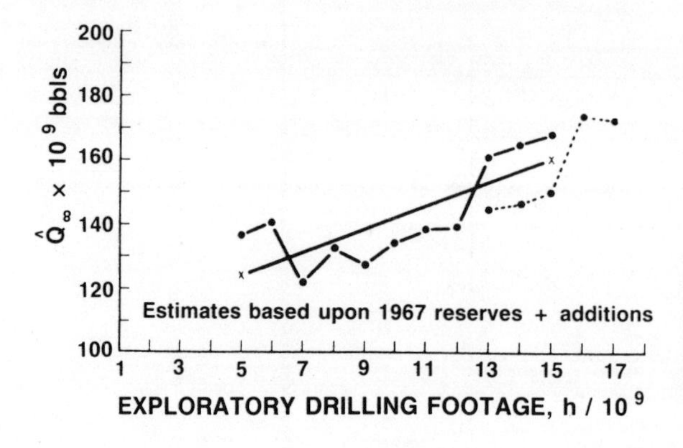

Figure 3.8. Successive estimates of Q_∞ by Hubbert's method as drilling is accumulated from 5×10^8 to 17×10^3 feet (modified from Harris, 1977).

somewhat grudgingly. As I remember, Root said that Harris had picked real nits.

Harris made this analysis while he was spending the summer in Washington, D.C., working at the National Science Foundation. During that summer, I tried to make him understand that when he looked at the data represented in Fig. 3.7, he was looking at a step function and not at a smooth decline curve. I discussed this idea from the perspective of what I had learned about the distributional form of the sizes of oil and gas fields. My point was that the decline curve had three phases. The initial phase was characterized by a high discovery rate but was of short duration. It was during this phase that the few largest fields were discovered. The second was a transition phase in which the discovery rate falls rapidly. The third phase covered a long period in which the discovery rate held at a low and nearly constant level. It was during this phase that thousands of small fields were discovered. Harris skeptically listened to my argument. His counterpoint was that a simple physical argument would not suffice because economic principles had to be evoked before a satisfactory argument explaining the pattern of oil and gas discovery rate could be made.

When I made my assertion to Harris, it was based on a combination of the then only partially documented idea that the largest fields are discovered very early during the exploration of a region and the associated idea that oil and gas field size distributions at all levels (that is, exploration plays to basins to countries and to the world) were very highly skewed in the right tail. Several years later, we had the data to document the three-stage pattern that discovery rate curves tend to follow throughout the exploration history of large regions (Root and Drew, 1979). In this study, we partitioned the discoveries made in the Permian basin into size classes and constructed discovery rate decline curves (Figs. 3.9 and 3.10). Because this basin accounted for 18 percent of the total crude oil discovered in the United States by 1974, we felt that it was large enough for us to make a generalization to the United States as a whole. Further data collection and analysis provided enough material to produce an atlas of

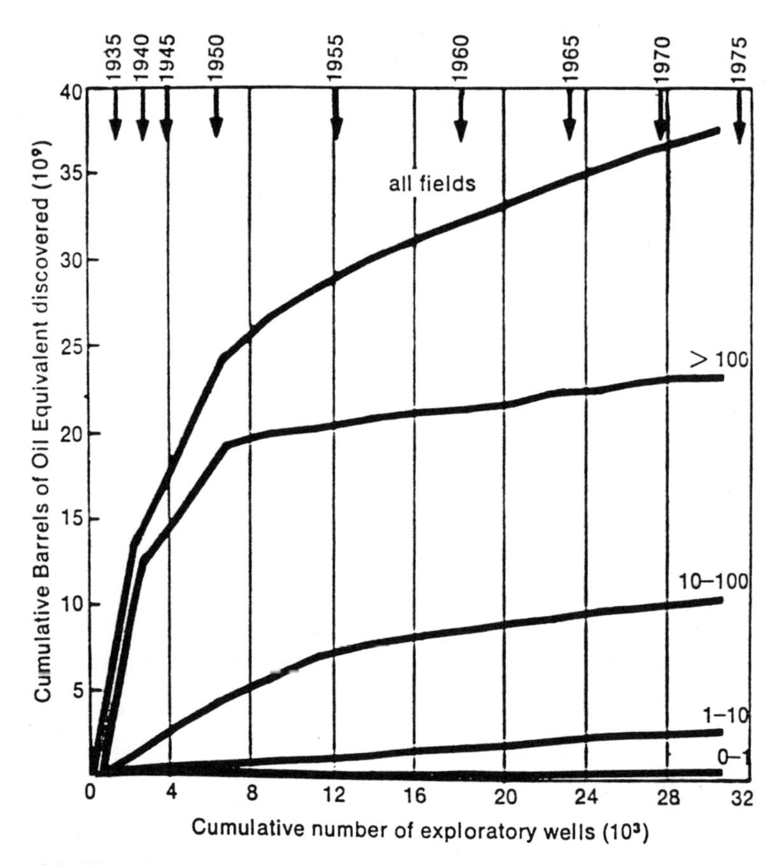

Figure 3.9. When the amount of petroleum discovered in the Permian basin is broken down into classes according to the sizes of the individual fields in which the petroleum occurs and amounts in these classes are plotted against the total cumulative number of exploratory wells, it becomes clear that the largest fields were discovered early. The black lines show the amounts of petroleum discovered in fields of size classes—those containing <1 million barrels of oil equivalent, 1–<10 million barrels of oil equivalent, 10–<100 million, and ≥100 million—at intervals in the excploration history of the basin. Note that although relatively little additional petroleum has been discovered in very large fields since 1950, petroleum from the largest class of fields still constitutes the major part of the total amount discovered (Root and Drew, 1979).

discovery rate curves that displayed the same three-step pattern in the discovery rate curves for 35 basins in the United States (Drew et al., 1983).

While M. King Hubbert was the main contributor and best-known scientist working on resource-assessment techniques at the USGS during the 1960s and early 1970s, there were other workers besides Vince McKelvey who must be mentioned. It is my opinion that the contributions of Al Zapp and Tom Hendricks were very significant. On many occasions, Hubbert credited Zapp with developing the calculus for the rate of discovery decline extrapolation method.

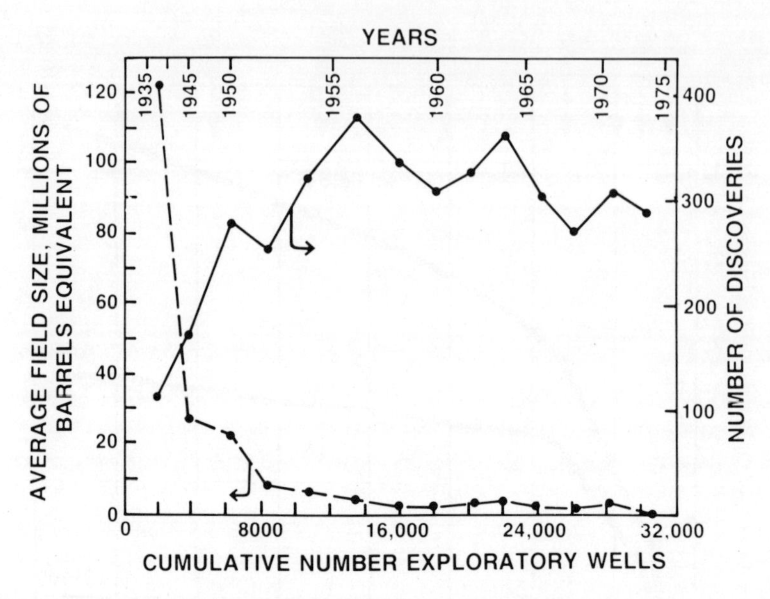

Figure 3.10. The drop in the average size of fields discovered after the first phases of exploration is vivid here. The period from 1921 to 1974 has been divided into 14 intervals, during each of which approximately 2000 exploratory wells were drilled. The intervals fall into three phases—an initial phase, ending by 1938, when the discovery rate was high; a transitional phase during which the discovery rate fell sharply; and a long phase, beginning in 1951, when the discovery rate was relatively low but stable. The actual number of fields discovered per drilling interval in the history of exploration of the Permian basin has fluctuated but has not shown a clear trend up or down. The success rate has not changed despite the fact that wells have become somewhat deeper over time (Root and Drew, 1979).

Any review of the development of techniques for estimating the volume of undiscovered oil and gas resources by USGS employees also must include the work of Tom Hendricks. I will be the first to say that the paper that Hendricks wrote in 1965 is difficult to read. It must be taken, however, as proof that he understood the concept of how to make a decline extrapolation forecast of ultimate productivity of U.S. crude oil. The nature and degree of the cross fertilization of ideas among Hubbert, Hendricks, Zapp, and the others at the USGS is not easy to determine, and, probably, it is only necessary to mention that by the early 1960s, there were a number of research scientists at the USGS working to develop resource-assessment techniques.

The field of petroleum-resource assessment evolved rapidly during the 1970s. McKelvey, Hubbert, and the other scientists who had focused their attention on the development and application of large-scale assessment techniques for petroleum resources had, by then, moved the field to a mature status. In a few words, they had accomplished the majority of what could be accomplished by using these techniques. The stage then expanded to include the development and use of assessment techniques that employed detailed geologic data and the statistical and mathematical analyses. These analyses explicitly modeled the relation between discovery and exploratory drilling and the order of discoveries. The distribution of field sizes became a critical consideration in many of these analyses. One of the best studies in this area was published in 1958 by J. J. Arps and T. G. Roberts but, for the most part, went unnoticed.

The volumetric assessment methods used by McKelvey and his coworkers at the USGS and by others elsewhere in the 1960s were still being used during the mid-1970s at the USGS by Bill Mallory. I liked Bill Mallory and believed that he made a heroic attempt to save this approach from extinction by constructing an enormous data base that included data on the volume of rock within each major stratigraphic unit in the entire United States that was favorable for the occurrence of oil and gas. He used the well history control file assembled by the Petroleum Information Corporation to compute a measure of the intensity of exploration. Mallory found that estimating the rock volume richness factor was a difficult task that needed to be studied further. In 1975, when he was involved with this research, the energy crisis was at its peak, and, instead of pouring additional research effort into further development of the volumetric approach, a new technique was chosen by the USGS.

This new technique was based on subjective judgment and became the basis for the USGS official assessment of the undiscovered oil and gas resources in 1975. For more than 10 years, this technique has been used by the USGS and the Canadian Geological Survey for large-scale assessments (McCrossan and Porter, 1973; Miller et al., 1975; Dolton et al., 1981; Masters et al., 1983; Procter et al., 1984; Mast et el., 1988). This technique has evolved rapidly from one that relied almost entirely on the application of expert judgment of geologic and geophysical data to one that relies on inputs from many sources. Discovery process modeling is now one of the sources, and it contributes information on the sizes of the oil and gas fields to be discovered in the future (Mast et al., 1988; Procter et al., 1988). This contribution and the associated analyses of field size distribution entered the mainstream of oil- and gas-resource assessment during the early 1980s (U.S. Geological Survey, 1980; Drew et al., 1982; Farmer, 1982; Farmer and Zaffarano, 1982; Energy Information Agency, 1983; Vidas and Duleep, 1984).

Although discovery process modeling came to life as an activity when Arps and Roberts published their paper in 1958, this technique was not used widely until nearly 20 years later; for example, my own involvement in this field began in 1972 with an attempt to empirically isolate the relation between wildcat drilling and discovery behavior inside an exploration play. From the experience I gained while working at Cities Service Oil Company, I realized that there was a bidirectional linkage between the act of drilling and discovery. Everyone knows that you have to drill wildcat wells to make discoveries, but it is not so obvious to the layperson that there is a large linkage effect between the size of recent discoveries and the number of wildcat wells drilled near and soon after these discoveries. In fact, most wildcat wells are drilled shortly after and somewhere near one or more large recent discoveries. Drilling a surge of wildcat wells behind large recent discoveries produces the cyclicity within an exploration play, which is one of the most important characteristics of a play. This cyclicity can be likened to the activity that goes on during a gold rush. I wanted to see if this cyclicity could be defined in a tight enough analytical form to be used to describe the exploration play as a fundamental unit of business activity and then to go on and use it as a basis for forecasting future rates of discovery of oil and gas in the United States.

During this same period of time, research efforts were also under way in universities and oil company research laboratories to study the behavior of exploration plays, discovery process modeling, and field size distributions. To cite a few of these, I mention the work of J. T. Ryan (University of Alberta), Gordon Kaufman (Massachusetts Institute of Technology), and David White and his colleagues (Exxon Production Research Company). Ryan's work was of particular interest to me because he had taken a nearly identical approach to analyze the exploration plays that had unfolded in the Western sedimentary basin in Canada as I had to analyze exploration behavior in the Powder River basin, Wyoming (Ryan, 1973a, b). Gordon Kaufman and his colleagues developed mathematical formulations for the behavior of exploration plays (Kaufman et al., 1975; Barouch and Kaufman, 1977). It is my recollection that Kaufman coined the phrase "discovery process modeling" for the activity in which we became involved. At Exxon Production Research Company, an intensive effort was under way to develop oil- and gas-assessment techniques that would blend geologic reasoning and statistical principles. The technique developed by the Exxon team was adopted in the mid-1970s by the Canadian Geological Survey (McCrossan, 1973; McCrossan and Porter, 1973). The assessment technique used by the USGS today is also similar. Although there were many members on this team, the names D. A. White, R. A. Baker, and H. M. Gehman appear most frequently (White et al., 1975; Gehman et al., 1975, 1980; White and Gehman, 1979; White, 1980; Baker et al., 1984). Although each of us began by working independently, we soon started talking to each other and adapting ideas from each other's work to the point that today many similarities exist in our combined collection of assessment techniques.

4

The Exploration Play

At the end of the last chapter, I mentioned the contribution made to the body of assessment techniques by the discovery process modelers who studied the behavior of exploration plays. The purpose of this chapter is to examine more closely the behavior of wildcat drilling and the discovery rates within exploration plays. The analysis of the behavior of exploration plays began with the splendid work of J. J. Arps and T. G. Roberts (1958) on the economics of exploratory and development drilling on the eastern flank of the Denver basin. Although the main focus of their study was on exploration and development economics, the behavior they modeled was that of an exploration play. In addition, this paper contained several insightful remarks about the nature of the parent size population of oil and gas fields.

Although I had read the Arps and Roberts paper sometime in the late 1960s, I did not fully recognize its importance. The premise on which they constructed their discovery process model was that, within specified field size classes, future discoveries would follow a declining exponential function of the number of wildcat wells drilled. I recall being puzzled by this formulation, but Arps was revered for his analytic creativity in the oil industry, and I suspect that is why I kept this paper in the bottom of my correspondence in-box for many years. Sometime in 1976, I wrote a computer program to calculate values for the expected rates of discovery specified by the Arps and Roberts model. I used a set of wildcat and discovery data from the Denver basin and made my first forecast using this discovery process model. I set up the model originally used by Arps and Roberts to forecast the number of discoveries to be made within each of the size classes and began the forecast as of the end of 1958. After I ran a forecast for the years 1959–1971, I was absolutely amazed when I checked the number of discoveries predicted by the Arps and Roberts model against the actual number of discoveries made—the results were nearly identical.

Before long, I realized that the predictive power of this model could be maximized if it were used to forecast future discovery rates within an exploration play. It is curious to note that Arps and Roberts applied their discovery model to the set of wildcat drilling and discovery data that later became known as the D-J Sandstone play of the Denver basin. Today, the data from this exploration play is considered to be the archetypal example of an exploration play. Arps and Roberts never said whether it was by accident or design that they chose this play as the test case for their model. Unfortunately, I never had the chance to talk to Arps about the discovery process model. Although I heard recently that Roberts was still alive and working for a bank in Dallas, I have never tried to contact him.

After the pioneering work of Arps and Roberts was published in 1958, no new analyses on the behavior of exploration plays were published until 1973, when J. T. Ryan wrote the first comprehensive analysis of the role of the exploration play phenomenon in petroleum-resource assessment. He investigated the initiation time sequence and the internal discovery rate behavior of the collection of exploration plays that had unfolded in Alberta, Canada (Ryan, 1973a, b). Ryan was first and foremost interested in explaining what appeared to him to be the erratic growth of crude oil discoveries in Alberta over the total time span of exploratory drilling (Fig. 4.1). The second task that he set for himself was to construct a model to forecast the future rate of discovery and to estimate the total recoverable crude oil in Alberta.

The illustration shown in Fig. 4.1 is a cumulative graph showing the growth in the total volume of crude oil discovered in Alberta. This graph is character-

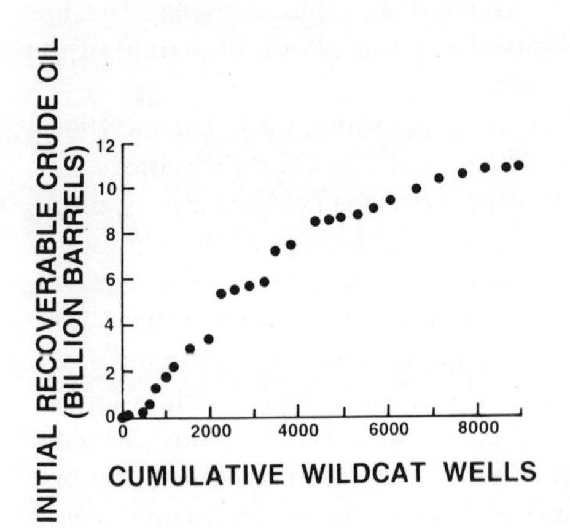

Figure 4.1. Growth of initial recoverable crude oil reserves in Alberta (Ryan, 1973a).

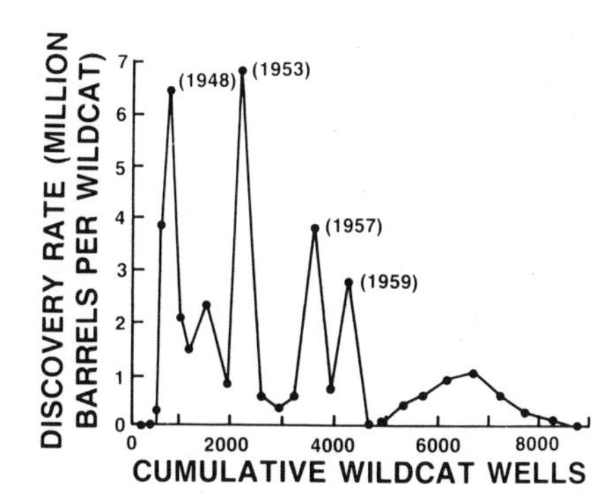

Figure 4.2. Discovery rate of recoverable crude oil in Alberta (Ryan, 1973a).

ized by sudden jumps followed by flat sections. The spikes in the graph shown in Fig. 4.2 are equivalent features in the incremental contribution to total discoveries. It is to these spikes that Ryan referred when he said that the growth of discovery of crude oil in Alberta is erratic. When Ryan categorized the oil fields according to the exploration play in which they were discovered, he found that the growth in total crude oil discoveries within each play was very regular (Fig. 4.3). To me, this conclusion stands out as his biggest contribution. For many years, graphs like those shown in Figs. 4.1 and 4.2 had been constructed and extrapolated by resource analysts for forecasting future rates of discovery. It was not until Ryan's analysis that the all-important contribution of the exploration play to the composition of such graphs was formally recognized.

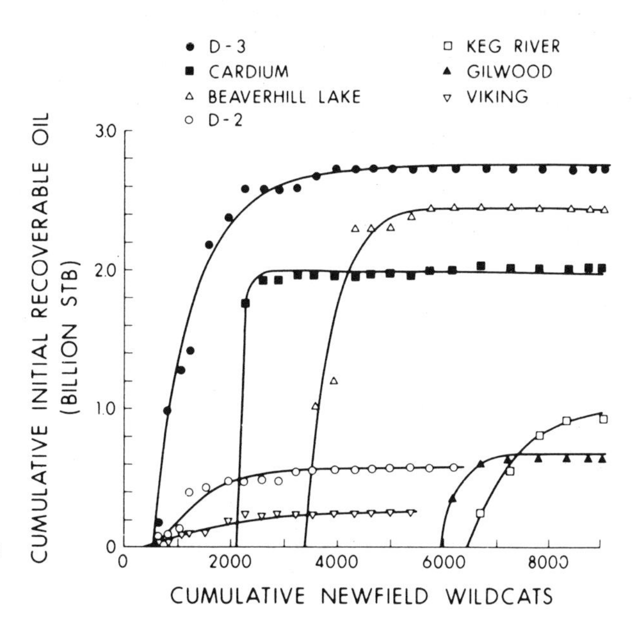

Figure 4.3. Recoverable oil reserve growth by play (Ryan, 1973a).

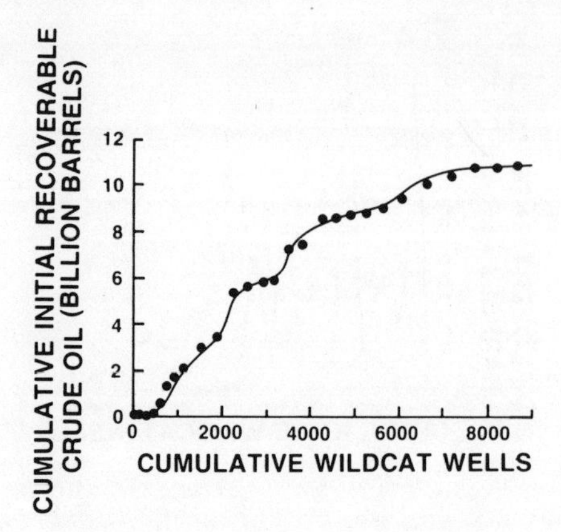

Figure 4.4. Growth of initial recoverable crude oil reserves (Ryan, 1973a).

Ryan's forecasting model was based on an idea that he termed a "knowledge function." He postulated that this function would grow from a value of zero (no knowledge at the start of a play) to a value of 1 (complete knowledge early in the development of a play). In a sense, this function is similar to the blocking or dummy variables used in econometric analyses to capture systematic shifts in a data series.

Ryan fit the crude oil discovery data series to his model and obtained good agreement (Figs. 4.4 and 4.5). This close fit of the model to the data is not a surprise because most of the degrees of freedom in the data were used in the process of fitting the data. Ryan was forthright in warning his readers about the risk involved in extrapolating such a model to forecast future discoveries (Ryan, 1973a, p. 233; 1973b, p. 245).

Ryan used his model to forecast the ultimate crude oil discoveries in Alberta and compared his forecast with similar published forecasts for Alberta (Table

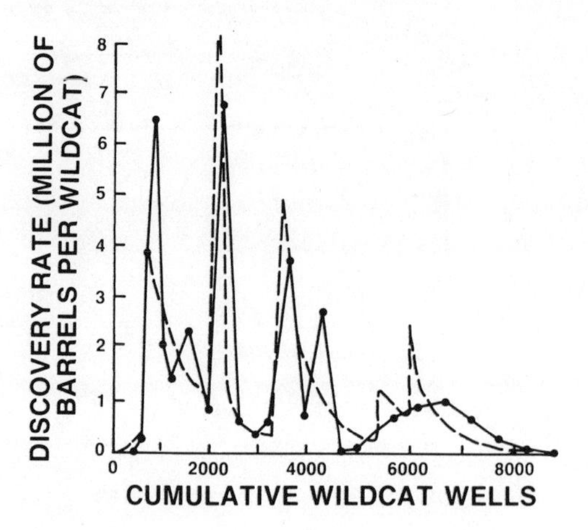

Figure 4.5. Discovery rate of recoverable crude oil (Ryan, 1973a).

Table 4.1. Forecasts of Undiscovered Crude Oil in Alberta, Canada

Estimator	Recoverable[a]	Oil in place[a]
Canadian Petroleum Association	25	75
Energy Resources Conservation Board	20	60
Van de Panne	17–18	51–54
Folinsbee	12–13	36–39
Ryan	12–13	35–39
McCrossan	16	48

[a]Billions of barrels.
Source: Ryan, 1973a.

4.1). In this comparison, he pointed out that his forecast did not include the possibility of discovering crude oil in a new exploration play. As Gordon Kaufman (1983, p. 247) pointed out in his thorough analysis of Ryan's work, "Ryan delegates this problem to the geologist." By avoiding the prediction of the startup of new exploration plays in Alberta, Ryan not only delegated the problem to the geologist, but he avoided the issue of attempting to predict the most important class of future events. He should take solace in the fact that this problem still remains to be solved.

At about the same time that Ryan was performing his analyses, I was involved in the study of the Minnelusa and the Muddy exploration plays of the Powder River basin, Wyoming. The original purpose of my work was to examine the exploration process from the point of view of an industrial organization analyst rather than to build a model to forecast future rates of discovery in the basin. Emphasis was placed on examining how and why the process worked the way it did. Several things interested me. Why did such explosive behavior in the rate of wildcat drilling occur when an exploration play began? Could this behavior be predicted?

I had formed an image of the exploration play from nothing more than reading the *Oil and Gas Journal* each week. Over the years, I watched the intensity of land leasing and wildcat drilling shift from basin to basin, from country to country, from onshore to offshore, from shallow to deep and back to shallow again, and sometimes to return to the same location abandoned many years before. To these observations, I added experience gained from working in the oil industry; for example, I remember talking to one of our landmen as he described the parcel of land he was frantically trying to acquire at that moment. When I stopped by a few weeks later, he was chasing after another land parcel. Along with these current action items, he was trying to work on long-term deals, which came in the form of the submittals of big parcels of land, usually from land promoters. Often, these submittals came in from other oil companies that wished to share the risk. I had also noticed that in each regional office there was usually a geologist who had been assigned the task of trying to develop a geologic concept that could make an exploration play. From these data, I concluded that there are two types of exploration activity—the frantic, short-term activity inside an ongoing play and the slower, more determined activity of synthesizing regional geologic data aimed at the goal of starting a new exploration play.

From these observations, I concluded that there were two types of wildcat drilling. I termed the drilling that went on inside of a play "cyclical wildcat drilling" and that was devoted to testing the ideas of where a new exploration play could be started "ambient wildcat drilling." I supported this conclusion with the idea that the exploration targets drilled inside of an exploration play are well known before the act of drilling a well. In other words, the petroleum zone of the play is well defined; for example, it might be a particular stratigraphic position in the clastic wedge or a group of piercement structures or some other well-defined geologic entity. This specific quality causes the exploration activity that goes on inside of a play to be sharply focused. The cyclical attributes of the leasing and drilling activities inside an ongoing exploration play are to be expected as the various exploration interests compete with each other to discover the oil and gas fields in the petroleum zone of the play.

The development of the geologic concepts and the synthesis of regional geologic data necessary to cause the drilling of wildcat wells at the start of a new exploration play are time-consuming tasks. Before a wildcat well, which is intended to start a new exploration play, is drilled, all levels of management fully participate in an extended period of discussion and justification because, unlike for an ongoing play, no tangible proof exists that oil and gas will be found. The reasons for drilling such wildcat wells are examined carefully because they are being driven by concepts whose connection to a future income is rather intangible. Often such a wildcat well reveals data that negate the years of work it took to justify its drilling. As a result, this type of wildcat drilling tends to be infrequent and also geographically diffuse. Therefore, I gave this type of wildcat drilling the name "ambient" in contrast to the wildcat drilling that goes on inside of a play and that is easily justified by the close association with the recent generation of income.

In the next few pages, I will present a demonstration of the differences in the character of ambient and cyclical wildcat drilling by using the 1950–1971 exploration history of the Powder River basin, Wyoming. This discussion is more technical than the preceding in that the parlance is, in large part, that of a statistical analyst.

In the Powder River basin during the 1950–1971 period, two major and several minor exploration plays unfolded. The first of these plays occurred in the Minnelusa Sandstone during the late 1950s and early 1960s, and the second, in the the Muddy Sandstone during the late 1960s and early 1970s. Minor plays occurred in the Dakota and the Parkman Sandstones during the middle 1950s. In total, 160 oil and gas fields were discovered by drilling 3691 exploratory wells (Drew, 1975a).

The ambient and cyclical components of exploratory drilling and discovery within this basin were isolated from the total drilling and discovery record by reducing the data set to a collection of half-degree cells (Fig. 4.6). Within each of these cells, the number of wildcat wells drilled each year was tabulated (Fig. 4.7). The total volume of petroleum discovered in each cell during each year is shown in Fig. 4.8. The number of oil and gas fields discovered each year and their depths are shown in Figs. 4.9 and 4.10, respectively.

To determine the ambient rate of wildcat drilling within each cell, a rule was established in which the observed level of wildcat drilling in each cell in a given year is classified as ambient if no significant discoveries were made in the cell or in any of the adjacent cells during that year or any of the three preceding years. For those years when the observed wildcat drilling rate in a cell was part of the surge attributable to an exploration play, the ambient wildcat drilling rate was estimated by using the mean ambient wildcat drilling rate for the cell.

The use of this allocation rule resulted in the partitioning of 1146 and 2545 wildcat wells into the ambient and cyclical phases of wildcat drilling, respectively, between 1950 and 1971. The wildcat drilling effort that was expended during the ambient phase (31.1 percent of the total), however, discovered 55.8 percent of the total petroleum found (371.1 million out of 665.5 million barrels). Inasmuch as 31.1 percent of the total wildcat drilling resulted in finding 55.8 percent of the total discoveries, the ambient component is 2.8 times as effective as the cyclical component.

The frequency distributions of the volume of petroleum contained in the fields discovered during both phases of exploration are shown in Fig. 4.11. Both of these distributions are bimodal in form. This bimodality is a consequence of the differences in the sizes of the fields discovered in the two major exploration plays that occurred in the study area. The fields forming the peaks in the 6.0–7.0 logarithmic unit range are predominantly in the Minnelusa Sandstone, whereas the peaks in the 4.0–5.0 logarithmic unit range are predominantly small single- or double-well fields in the Muddy Sandstone. The general form of these two distributions also is somewhat different. The distribution describing the pattern of field discovery during the ambient phase of exploration is nearly uniform in shape, whereas that of the cyclical phase declines as field size increases; the latter is a consequence of the numerous small fields discovered within the Muddy and the Minnelusa plays as they progressed. It is my judgment that most of these smaller fields are too small to start a play but are large enough to keep it going once it is underway.

Although the average return, in barrels of petroleum discovered, on a wildcat well drilled during the ambient phase of exploration was 2.8 times higher than that on a wildcat well drilled during the cyclical phase, it was achieved at a higher drilling risk. The risk of failure was higher during the ambient phase, which had only 3.5 chances of success per 100 wildcat wells drilled. During the cyclical phase, 5 wildcats per 100 were successful. It may be concluded that the

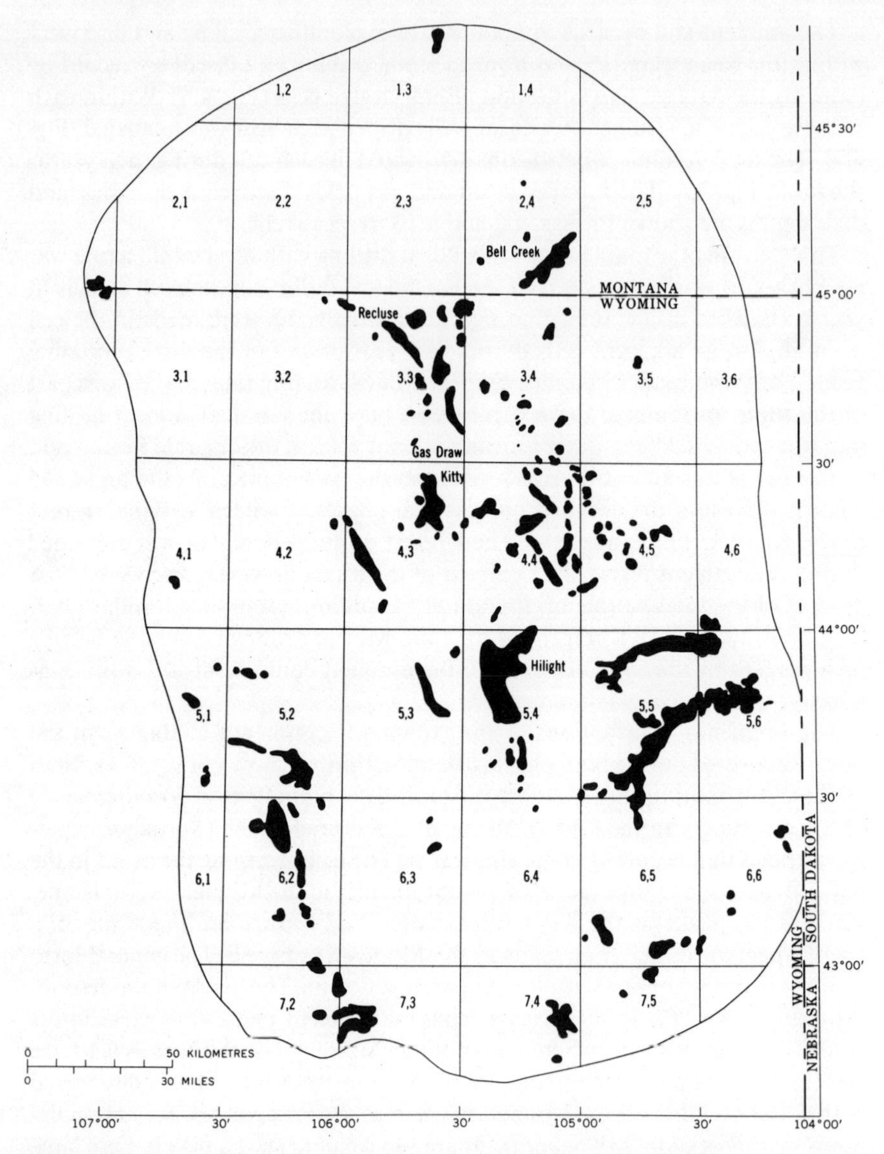

Figure 4.6. Location of petroleum deposits and cells in the Powder River basin (Drew, 1975a).

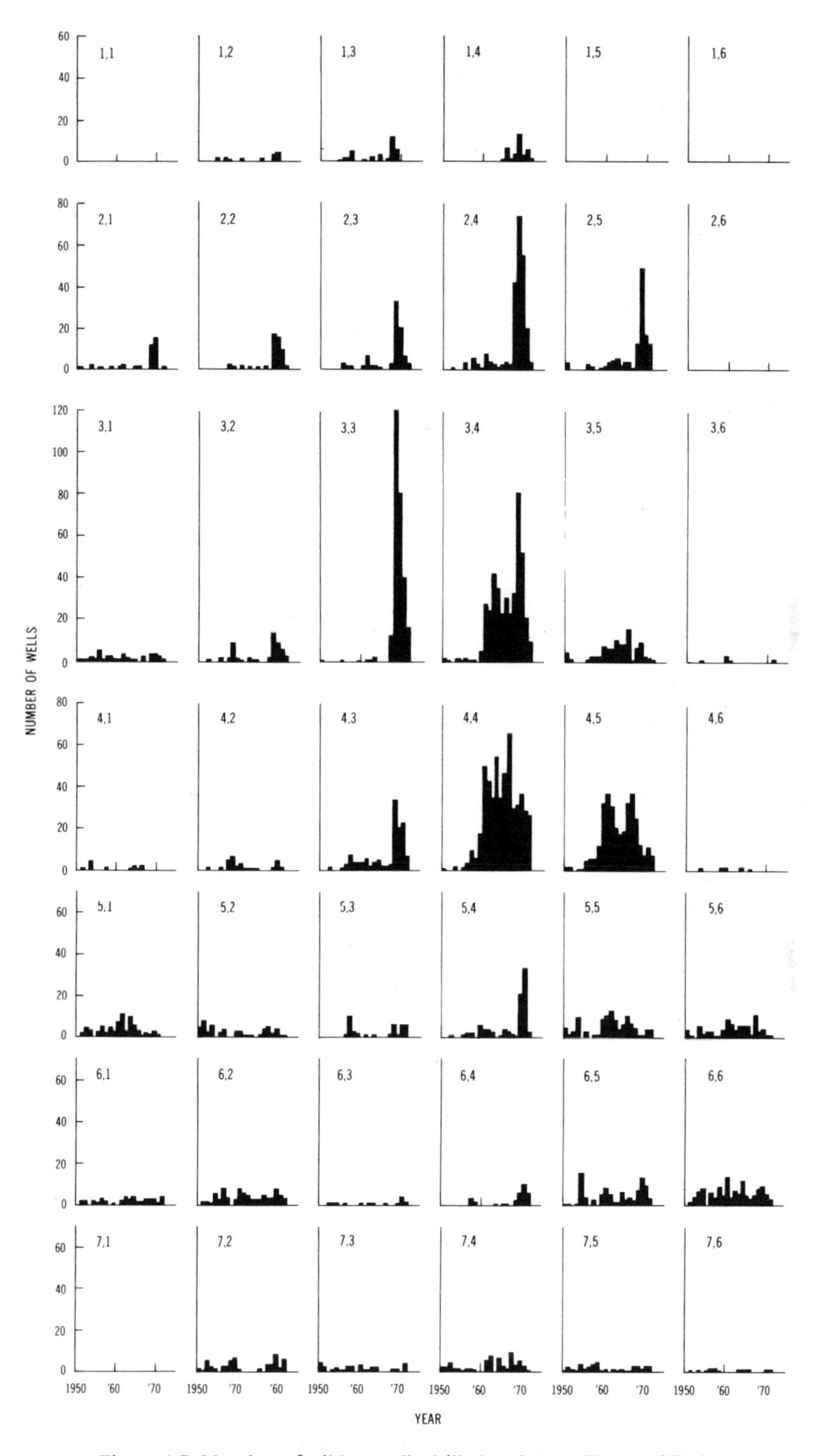

Figure 4.7. Number of wildcat wells drilled each year (Drew, 1975a).

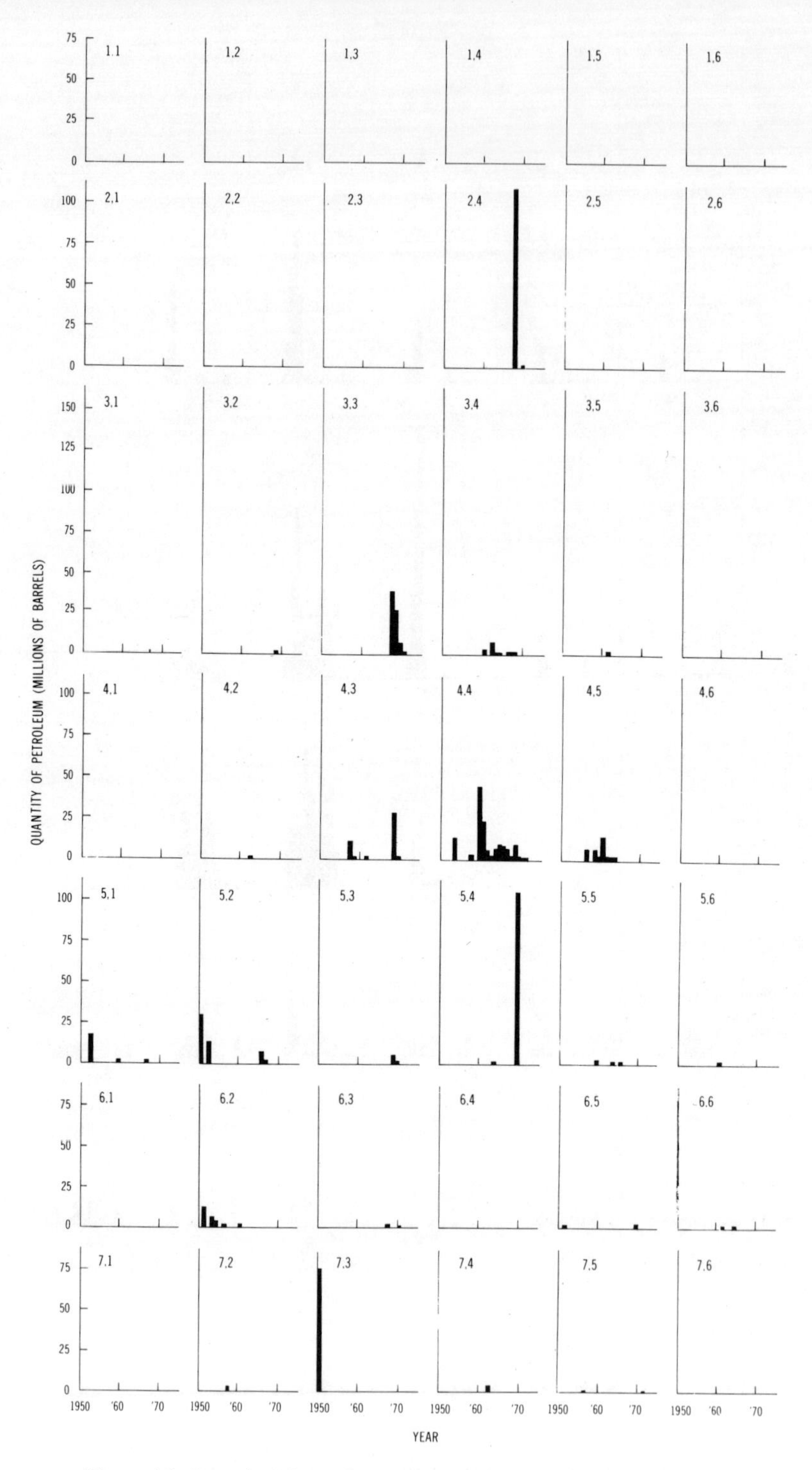

Figure 4.8. Quantity of petroleum discovered each year (Drew, 1975a).

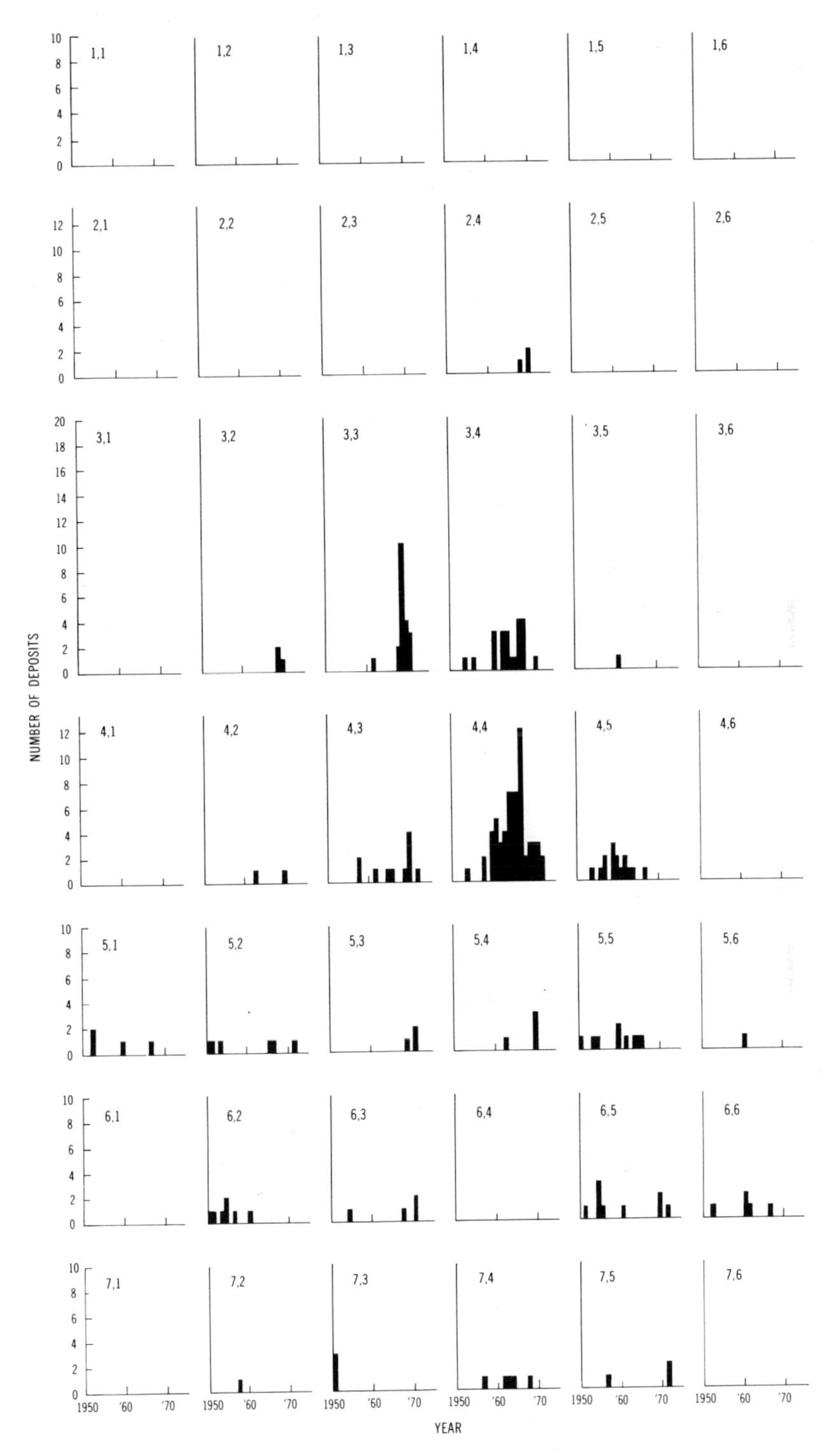

Figure 4.9. Number of oil and gas fields discovered each year (Drew, 1975a).

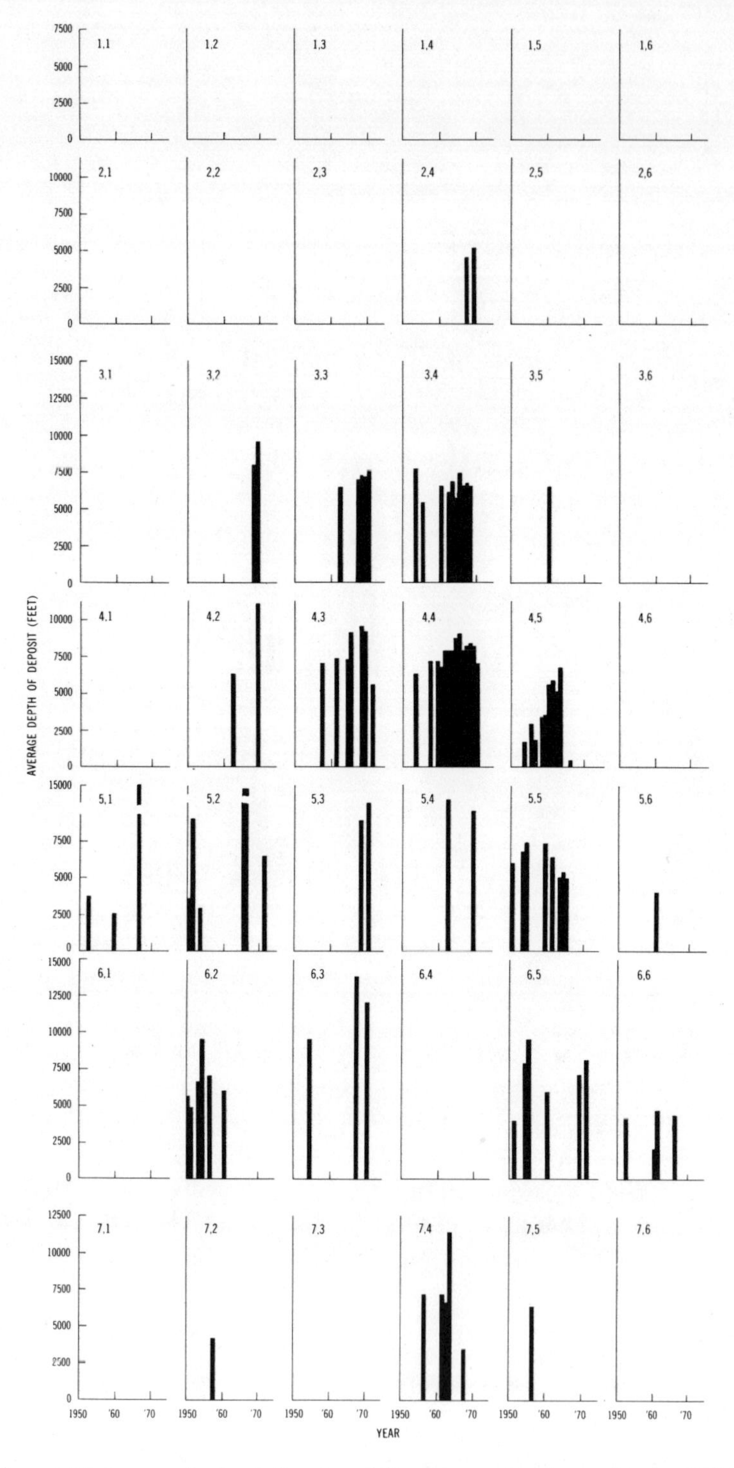

Figure 4.10. Average depth of oil and gas fields discovered each year (Drew, 1975a).

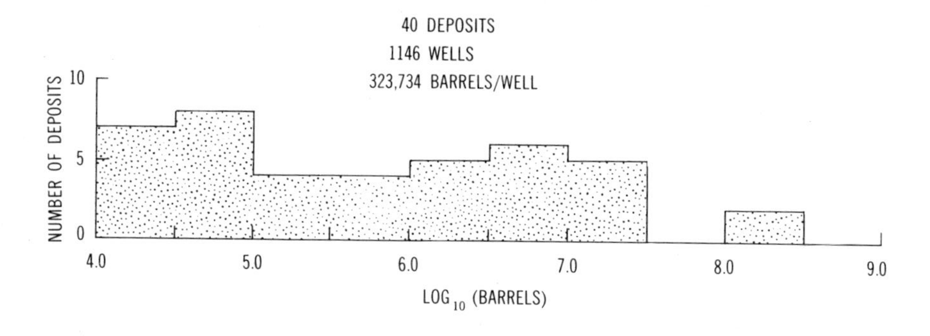

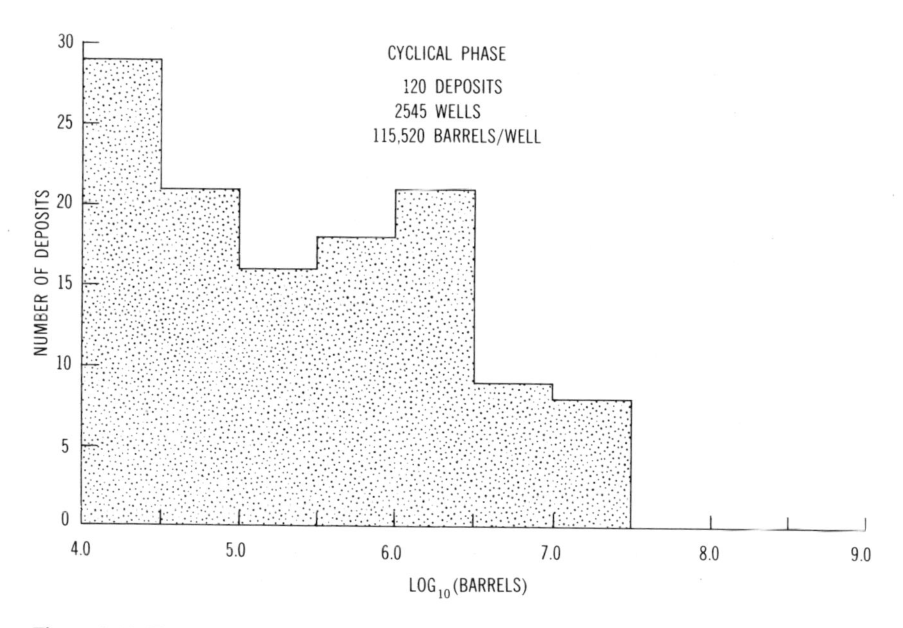

Figure 4.11. Frequency distributions of the volume of petroleum discovered in the ambient and cyclical phases of exploration (Drew, 1975a).

exploration play is a risk-reduction phenomenon in which the average size of discovery is traded off (nearly three to one in the Powder River basin example) for an increase in the probability of discovery (43 percent).

The behavior of the discovery success ratio within the Minnelusa and the Muddy exploration plays was examined in detail in a follow-up study (Drew, 1975b). This analysis was confined to that part of the Powder River basin in which these two plays upfolded (Fig. 4.12). The discovery success ratio for this region during the period 1957–1972 oscillated as these two exploration plays went through their cycles (Fig. 4.13).

During the initial phases of the Minnelusa and the Muddy plays, the success

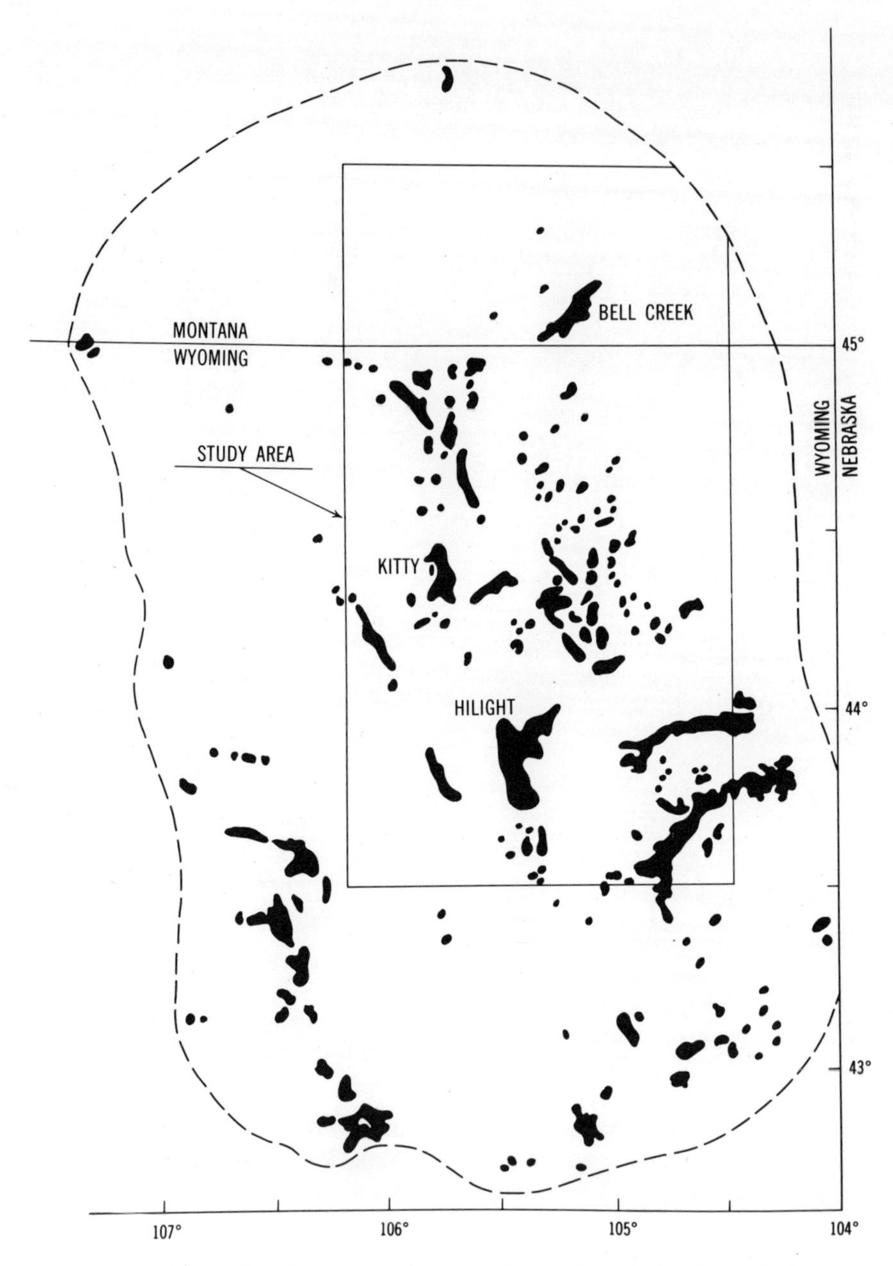

Figure 4.12. Location of study area within the Powder River basin (Drew, 1975b).

ratio fell sharply (Fig. 4.13). The decline in the success ratio that occurred between 1957 and 1958 coincides with the start of the Minnelusa play, and the decline between 1967 and 1968 coincides with the start of the Muddy play. These declines were then followed by a gradual recovery. In 1968, when the Muddy exploration play was at its peak, only 4.6 wildcats per 100 were successful. In 1969, the success ratio rose to 5.9 wildcat wells per 100, and it increased further in 1970 when 6.3 wildcat wells per 100 were successful. During 1971, the success ratio improved further to 8.1 wildcats per 100. During the 1958–1965 period, when the Minnelusa play was active, the success ratio curve followed a similar pattern (Fig. 4.13).

The pattern exhibited by this discovery success ratio curve can be interpreted as a type of learning curve. The initial phases of the Minnelusa (1958–1960) and the Muddy (1967–1968) exploration plays were characterized by a surge in the wildcat drilling rate (Fig. 4.14). This surge is typical of the manner in which exploration operators react to the news that a large oil and gas field has been discovered in a stratigraphic unit or other geologic structure not previously thought, with any certainty, to contain such a large field. The surge of

Figure 4.13. Relation between discovery success ratio and time in the Minnelusa and Muddy plays in the Powder River basin (Drew, 1975b).

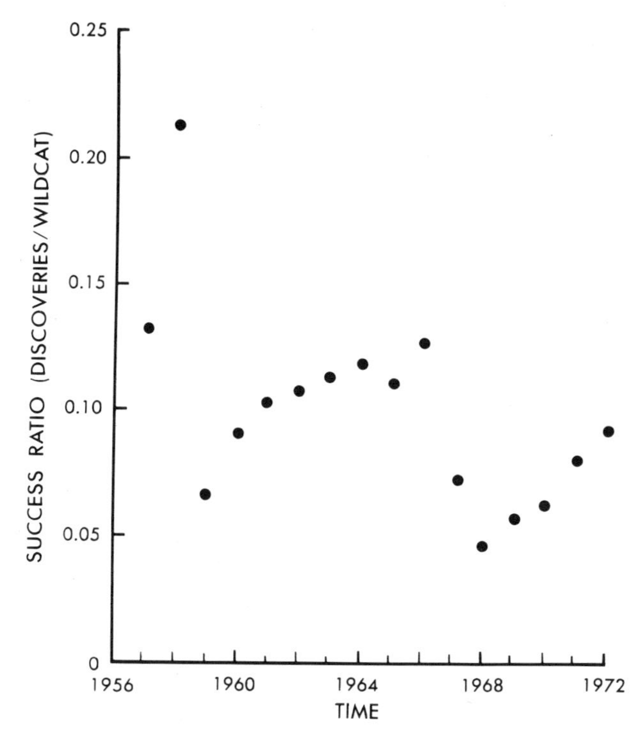

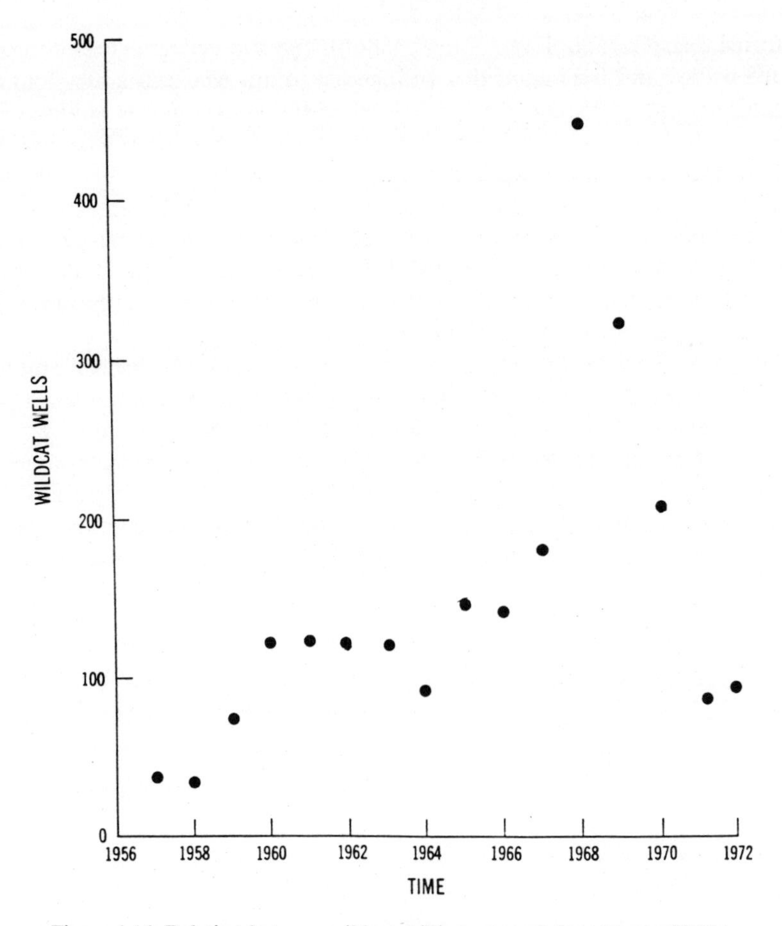

Figure 4.14. Relation between wildcat drilling rate and time (Drew, 1975b).

wildcat drilling at the start of an exploration play is the action taken to revise this expectation upward. This is the basis for the cyclical pattern of behavior, or gold rush phenomenon, of the exploration play.

Another explanation can be offered to explain the surge of wildcat drilling that occurs at the start of these exploration plays. During the initial phases of these plays, the enthusiasm of the exploration operators was in such a heightened state that, in their haste to test the Minnelusa and the Muddy stratigraphic units for additional large fields, many poor-quality exploration prospects were drilled. Consequently, the discovery success ratio fell sharply. As both of these plays matured and went into decline, the operators used the steadily increasing volume of geologic information to drill, on average, better and better quality prospects, at least in terms of the discovery success ratio.

By using the data displayed in Figs. 4.6–4.10, we can isolate a group of spikes in the wildcat drilling record that go together to describe how an exploration unfolds in time and space; for example, the Muddy exploration play was initiated by the discovery of the giant Bell Creek field located in cell 2,4 (Fig. 4.6). After the discovery of this field, which contained over 150 million barrels of reserves, the oil industry reacted, and the Muddy exploration play began. To give a mental image of the magnitude and the character of this reaction as expressed by the surge in the wildcat drilling rate, I like to use the analogy of how the energy from a seismic event is attenuated as it travels outward from its source. This analogy is useful in describing, for example, the sizes of the spikes in the wildcat drilling rates for cells 2,1–2,5 during the 1967–1970 period (Fig. 4.7). Note that the largest number of wildcat wells were drilled in cell 2,4 during 1968, which was the year after the Bell Creek field was discovered in that cell. Also note that the heights of the corresponding spikes in cell 2,3 going westward to cell 2,1 decline steadily. Eastward in cell 2,5, the height of the corresponding spike is smaller than in cell 2,4. Note that in this tier of cells the only cell where discoveries were made is cell 2,4. It can be concluded that the discovery of the Bell Creek field and, perhaps, several smaller fields, which were discovered shortly thereafter to the south located in cells 3,3, 3,4, 4,3, and 4,4, caused a wave of enthusiasm to spread out across the central and northern portions of the basin, a wave that took several years to die down and return to an ambient state. The data in Fig. 4.8 display the volume of petroleum discovery by year and show that the large spike in cell 2,4 in 1967 (the discovery of Bell Creek) dominates the discovery graphs for all cells in the central and northern portions of the basin.

Further inspection of the data displayed in Figs. 4.7–4.10 revealed that the wave of enthusiasm generated by the discovery of the Bell Creek field was amplified by smaller waves of enthusiasm created by numerous smaller discoveries made in 1968 and 1969 in cells 3,3, 3,4, 4,3, and 4,4. The linkage between the discovery of oil and gas fields and the subsequent drilling of wildcat wells was modeled by using regression analysis.

The regression model specified a set of equations that explained the effect of a discovery on the subsequent levels of wildcat drilling in the cell in which it was discovered and also in the adjacent cells. The effectiveness of this model in predicting wildcat drilling behavior is displayed in Figs. 4.15–4.18, in which the actual and predicted levels of wildcat drilling are displayed for four cells that experienced high levels of activity.

During 1968, the largest single increase in the level of wildcat drilling occurred within cell 3,3 (Figs. 4.7 and 4.15). In 1967, only 13 wildcat wells were drilled in this cell, whereas 121 wildcat wells were drilled during the following year. The predicted level of wildcat drilling in this cell during the 1967–1971 period is estimated for the discovery of 12 oil and gas fields within the cell and for 11 fields discovered in three of the four directly adjacent cells. The predicted

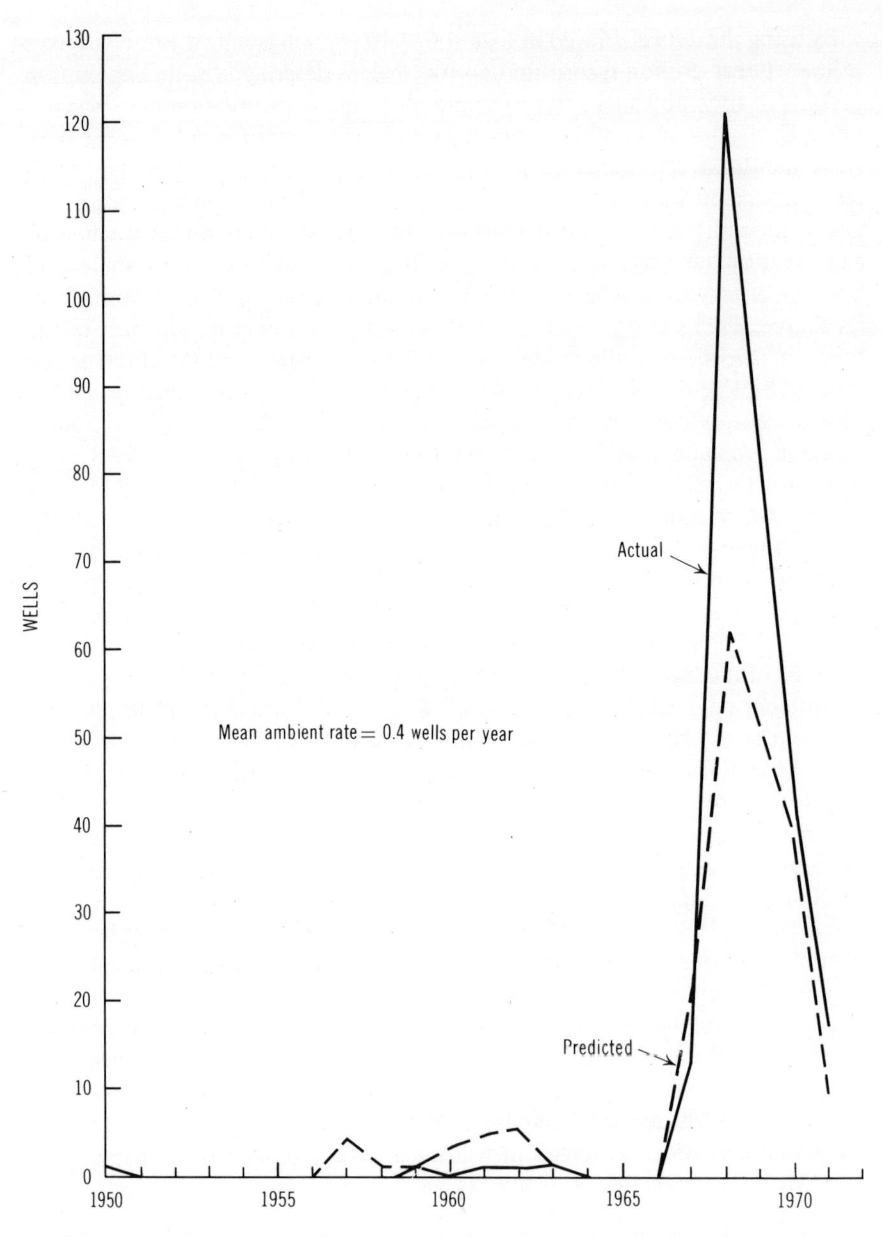

Figure 4.15. Actual and predicted rates of wildcat drilling in cell 3,3 (cyclical only).

levels are close to the actual levels, except in 1968 when the Muddy play was at its peak (Fig. 4.15). During this year, twice as many wildcat wells were drilled as the regression model attributes to the effect of recent nearby discoveries.

The variation in the level of wildcat drilling in cell 3,4 (Fig. 4.16) is more complex than that in cell 3,3 (Fig. 4.15). Between 1960 and 1971, only nine oil and gas fields, none of which contained more that 6 million barrels of producible petroleum, were discovered in this cell, although 425 wildcat wells were drilled. These wildcat wells were drilled mainly as a result of the discovery of the large volume of petroleum contained in the 45 fields discovered during this time period in the adjacent cells. The manner in which the model predicted the wildcat drilling rate within two additional cells where the discoveries are

Figure 4.16. Actual and predicted rates of wildcat drilling in cell 3,4 (cyclical only).

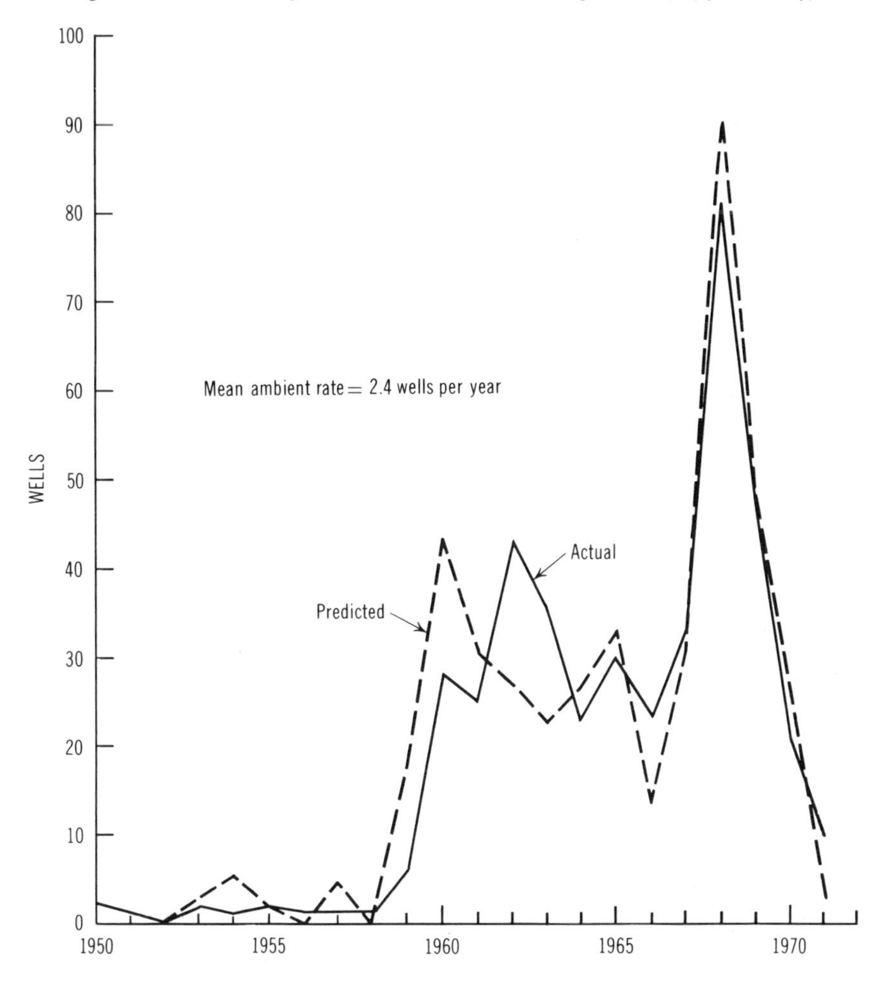

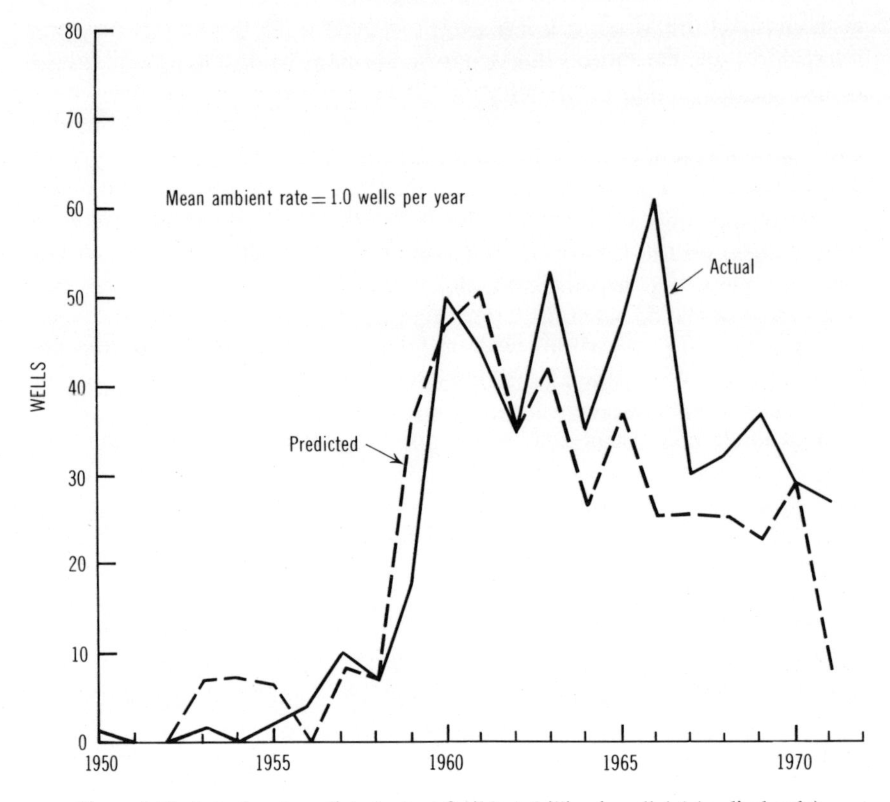

Figure 4.17. Actual and predicted rates of wildcat drilling in cell 4,4 (cyclical only).

more balanced between being in the cells and in adjacent cells is displayed in Figs. 4.17 and 4.18. In cell 4,4, the model modestly underpredicts the actual level of wildcat drilling as compared to cell 4,5, where the model modestly overpredicts the actual level of wildcat drilling.

The association between the volume of petroleum and the number of fields discovered and the number of wildcat wells drilled was examined by using these same data (Drew, 1975b). This analysis revealed that the number of discoveries made per year was nonlinearly associated with the number of wildcat wells drilled (Fig. 4.19) and that the volume of petroleum discovered varied independently of the wildcat drilling rate.

The nonlinearity in the relation between the number of fields discovered and the number of wildcat wells drilled could be interpreted as a diminishing return of the number of discoveries per wildcat well (Fig. 4.19). However, when the graph was annotated by and reference made to the analysis of the data shown in Fig. 4.19, it was obvious that this nonlinearity was mainly caused by the two data points from 1968 and 1969 when the Muddy play was at its peak. Because of the haste of the oil companies to test prospects in the Muddy stratigraphic interval in 1968, the success ratio fell to its lowest point during the 1957–1972 period. In 1969, the enthusiasm for the Muddy play slackened, and the success ratio began to rise as fewer wells were drilled into, on average, better quality prospects. Disregarding these two data points would produce a graph showing a constant rate of discovery as a function of the wildcat drilling rate. From this analysis, I concluded that more wildcat wells drilled per year in an exploration play meant more discoveries but not a systematic increase in, say, the volume of petroleum discovered per year.

Where does such a conclusion leave us when we examine the rate at which oil and gas are discovered in large regions, such as the Permian basin or the conterminous United States. In studies focused at these scales, we must relegate the exploration play to the world of small-scale phenomena. In large regions, such as the Permian basin, the rates of discovery are regular. Why should this happen? It seems as though the collective behavior of drilling and discovery of a group of exploration plays, which are initiated at different points in time, would be a prescription to produce an aggregate chaotic behavior; for example, we observed that fields in the 0.76-million- to 1.52-million-barrels-of-oil-equivalent size class in the Permian basin were discovered at the rate of 17 fields per 1000 exploratory wells drilled between 1921 and 1974 (U.S. Geological Survey, 1980). This is not a rough average, but a nearly constant rate; that is, every time another 1000 exploratory wells were drilled, almost exactly another 17

Figure 4.18. Actual and predicted rates of wildcat drilling in cell 4,5 (cyclical only).

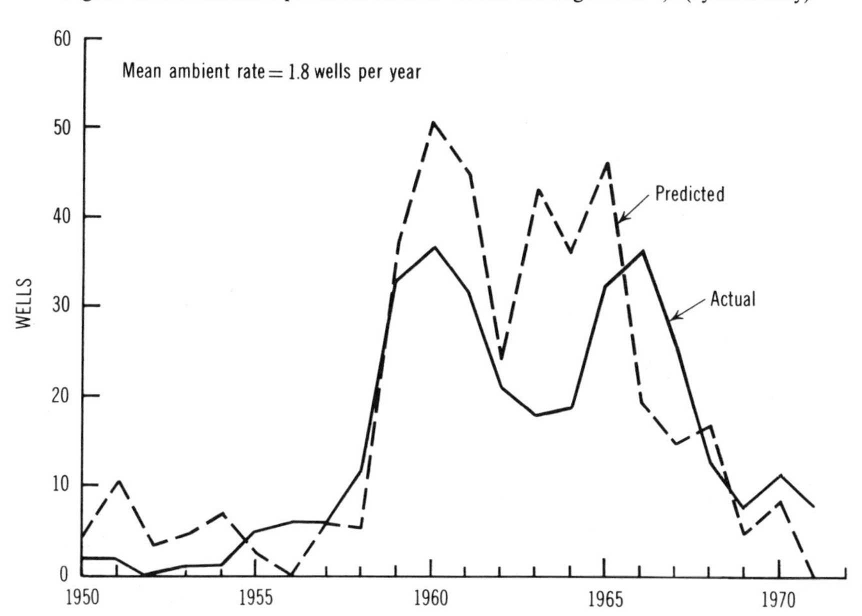

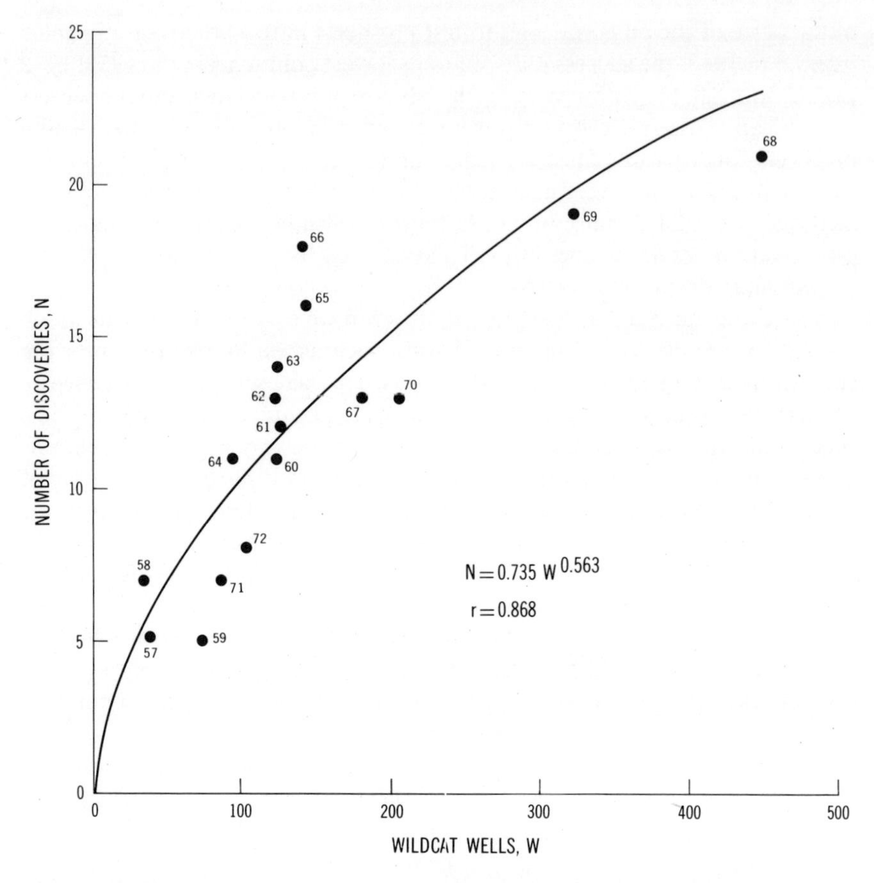

Figure 4.19. Relation between deposit discovery rate and wildcat drilling rate, data points annotated by year (numbers next to data points) (Drew, 1975b).

fields of this size were discovered. I began to understand when I reexamined Ryan's work in Alberta.

Ryan's incremental discovery rate graph for Alberta (Fig. 4.2) showed me that the exploration plays containing the larger volume of crude oil (the largest spikes) were initiated earlier in the exploration history than were the plays that produced the smaller volumes. The highest points of these spikes correspond exactly to the beginning of the exploration plays and the peak discovery rate. At these points, the largest fields have just been discovered. Also note that the area under each spike is the total amount of petroleum discovered within each play. These data led me to tentatively conclude that when a region is large enough, a number of exploration plays may unfold in it and that the pulselike behavior of the individual exploration plays will be smoothed as smaller and smaller exploration plays are initiated over time.

It also may be reasonable to expect that there is a regular pattern to the initiation of new exploration plays in large regions. When we solve the data integration problems that exist in disaggregating the cumulative drilling histories of large regions into exploration plays, we may find, for example, that the initiation sequence of new plays on a scale of cumulative wildcat wells is distributed as a Poisson point process in which the λ parameter of this distribution increases according to a specified function of cumulative wildcat wells. If we can show that the volume of oil and gas occurring within each play will decline systematically over time, it will follow that the size of the largest fields discovered in each play must decline because there is a correlation between the amount of oil and gas occurring in the largest field in a play and the total amount of oil and gas occurring in the play. Of course, it could be said that given the two scales of behavior mentioned above, one of which is a temporal composition of the other, the regularity at the larger scale is proof positive of a systematic composition. The empirical scientist wishes for the better state of affairs where he can trace the decomposition of an aggregate phenomenon into its component parts.

My study of the behavior of exploration plays continued with an analysis of the actions of the exploration firms that drilled the wildcat wells. I became interested in the structure of the exploration industry from the point of view of determining the roles played by the different types of exploration firms. My starting point was the belief that the exploration for crude oil and natural gas in the United States and elsewhere is not carried out by a monolith as is often portrayed by the press. Instead, there are distinct groups of firms ranging in size from single individuals to the major integrated oil companies, each of which uses a different exploration strategy (Drew, 1980; Drew and Attanasi, 1982).

Data from the D-J Sandstone play of the Denver basin covering the 1949–1971 period were used in this study. Part of the reason for examining the behavior of exploration operators was to contribute to the argument that was raging during the middle and late 1970s as to whether the domestic supply of petroleum could be increased by the divestiture and regionalization of the major integrated oil companies. This argument went along the lines of suggesting that the public was being victimized by the greed and wastefulness of the major oil companies. In a few words, petroleum products were not being offered to the public at competitive prices. The suggested solution was to break up the major firms into smaller ones that would compete more effectively with each other. I looked specifically at the notion put forth by industry spokespersons that there was a symbiotic relation between the major integrated firms and the other types of exploration firms and that it was in the interest of the nation as a whole to preserve it.

The exploration firms active in the Denver basin were classified in four groups. The first group consisted of the 20 major firms that were vertically integrated and active at all stages of the petroleum industry—from exploration and production on through transportation, refining, and product marketing. The second group consisted of 39 large, independent companies that were principally involved in exploration and production. The drilling contractors were classified as the third group of firms, which included only operators whose principal business is the sale of contract drilling services to the petroleum industry. The drilling contractors became the operators of record on many wildcat wells because they often explored as a sideline, and, in many cases, they participated in complicated financial dealings to get many wildcat wells drilled. The fourth group, the "small independents," consisted of more than 3000 individuals and small firms. Rarely did any member of this group drill more than 30 wells in any year.

Inspection of the time profile of discovery and wildcat drilling for the four classes of firms reveals that, by the late 1950s, the major firms had essentially stopped drilling wildcat wells in the Denver basin (Fig. 4.20). The major firms

Figure 4.20. Yearly aggregate rate of discovery of petroleum and wildcat drilling rates, by type of firm, in the Denver basin (Drew, 1980).

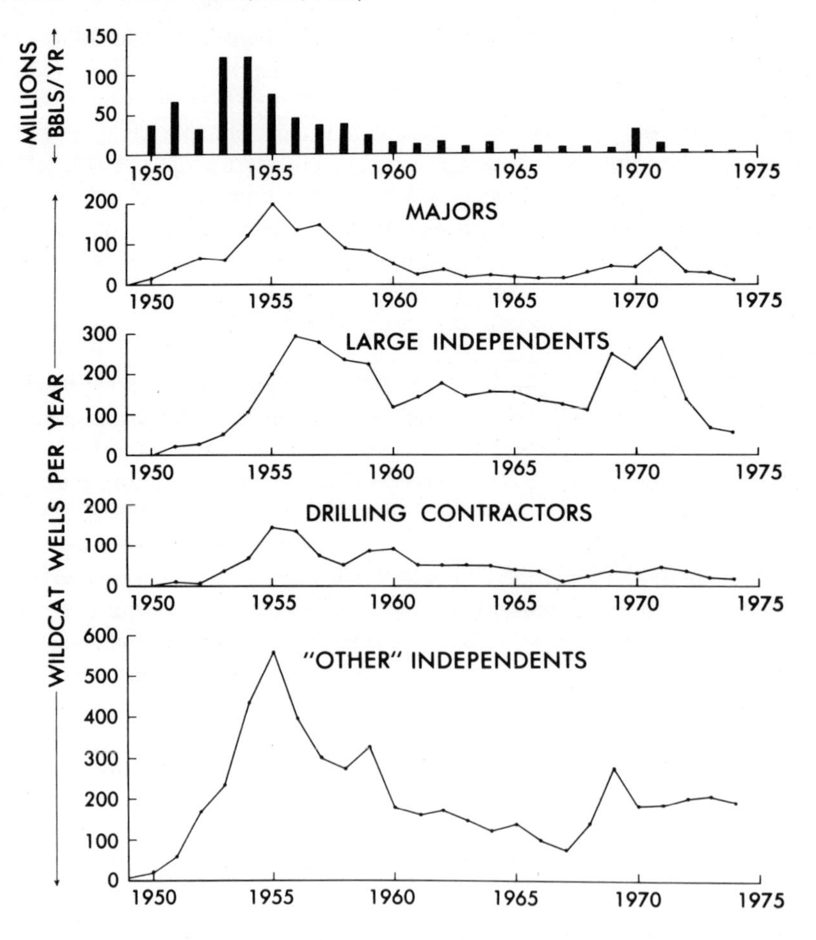

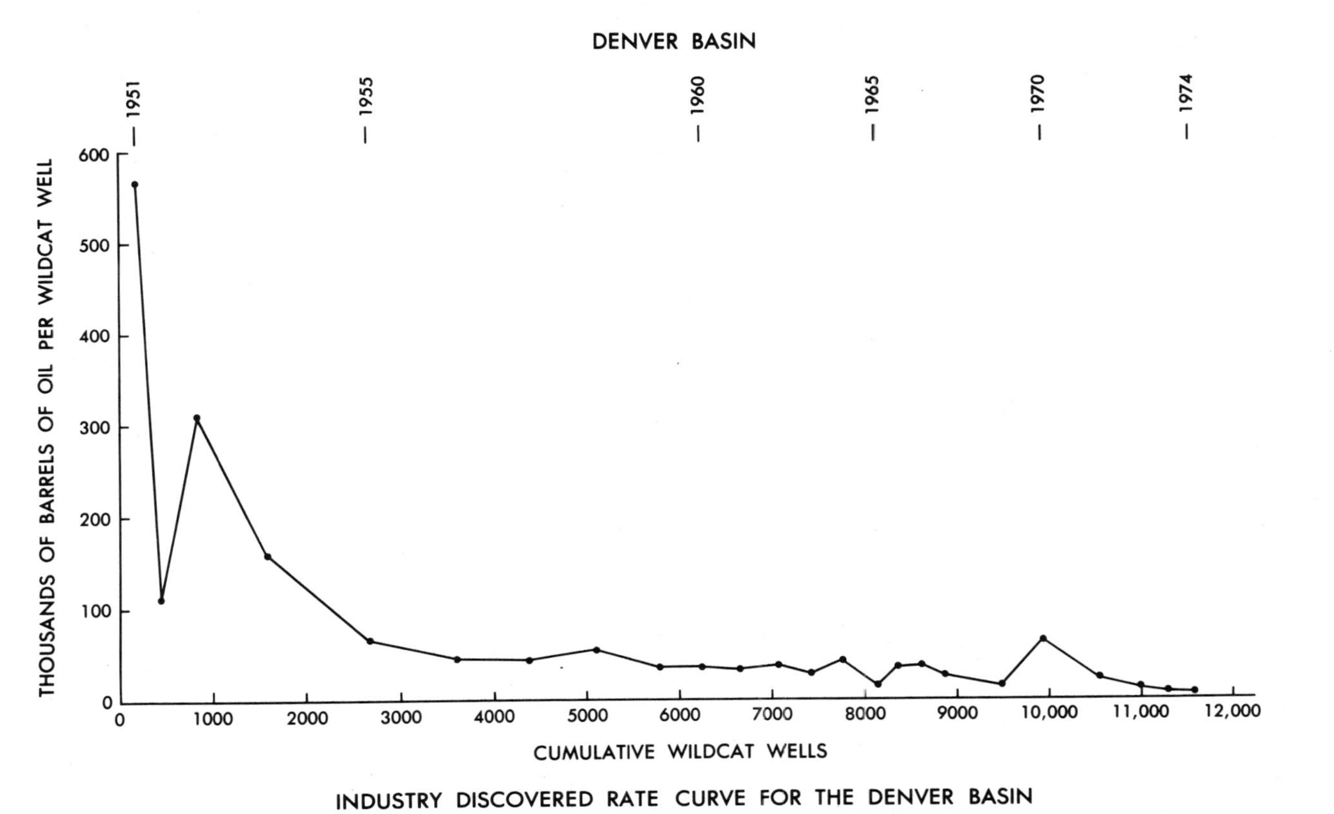

Figure 4.21. Graph showing the wildcat well discovery rate (barrels per wildcat well) for petroleum in the Denver basin during the 1949–1974 period (Drew and Attanasi, 1980).

left the D-J Sandstone play because the rate of discovery had fallen off sharply after 1953 (Fig. 4.21). By this time, the majority of the larger oil and gas fields had been discovered in the D-J Sandstone exploration play.

The number of wildcat wells drilled by the majors closely tracks the total volume of discoveries per year when wildcat drilling is lagged by two years. One of the conclusions that can be drawn from this result is that it takes that long for them to sense that an exploration play has peaked. The same sort of result had been found earlier in the analysis of the Minnelusa and the Muddy Formation plays in the Powder River basin. To a lesser degree, the drilling rates for each of the other classes of firms shows the same lag effect. Even though the rate of discovery had fallen to a rather low level (about 50,000 barrels per wildcat well) by 1955, the firms in these classes still continued to drill more than 200 wildcat wells per year throughout the 1960s (Figs. 4.20 and 4.21). Surely, this level of drilling activity alone indicates that the major firms did not prohibit the smaller firms from exploring in this basin.

The strategy that seems to have been followed by the majors was to assemble and hold large blocks of acreage over the long term. This allows them to systematically evaluate and select for drilling the higher quality acreage and then, in turn, to farm out the poorer quality acreage to other firms. The results of this strategy are displayed in Table 4.2. As a class, the major integrated firms discovered 206.9 million barrels of petroleum (28 percent of the total) in drilling 1475 wildcat wells, 144 of which were successful. They discovered 140,300 barrels per wildcat well, more than twice the amount discovered per well by any other group. Their wildcat success ratio was also the highest at 0.098 field discovered per wildcat drilled.

Most of the petroleum discovered between 1949 and 1974 in the Denver basin (72 percent) was discovered by the three other types of firms. Although the smaller firms drilled most of the wildcat wells and discovered most of the petroleum, their land positions were such that they were not able to obtain a

Table 4.2. Average Discovery Success and Volumes of Petroleum Discovered by Operator Class from 1949 to 1974 in the Denver Basin

	Large majors	Large independents	Drilling contractors	Small independents	Total
Number of successful wildcats	144	275	84	406	909
Number of unsuccessful wildcats	1331	3484	1228	4614	10,657
Wildcat-well success ratio	0.098	0.073	0.064	0.081	0.079
Petroleum discovered[a]	206.9	185.6	80.8	268.7	742.0
Petroleum discovered per wildcat well drilled[b]	140.3	49.4	61.6	53.5	64.2

[a]In million barrels of producible oil.

[b]In thousand barrels of producible oil.

Source: Petroleum Information, Inc. (1975).

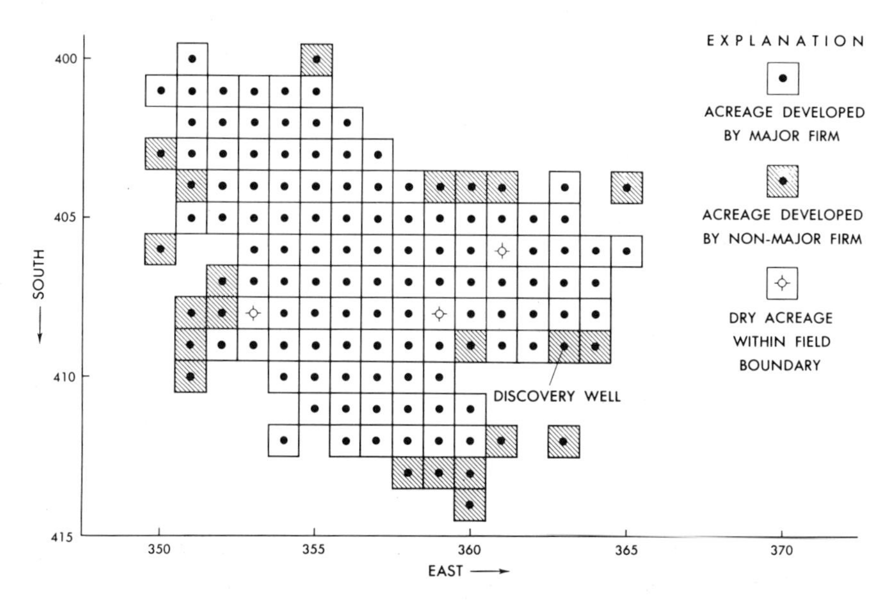

Figure 4.22. Map of acreage ownership within the Sloss field. The spacing is one well per 40 acres (Drew and Attanasi, 1980).

proportional share of the reserves that their wildcat wells discovered. This situation is illustrated in Fig. 4.22, where the locations of the discovery and the development wells drilled by type of firm in the Sloss field are shown. This field, which is one of the largest in the Denver basin, was discovered by an independent firm at a site that ultimately proved to be on the edge of the productive limit of the field. Note, however, that the major firms held or acquired a land position that allowed them to drill in most of the good drilling locations in the field. Also note that the margins of the field were drilled by independent firms. These sites almost always are inferior in terms of their reserves and productivity.

Field maps for the 20 largest fields show that they have the same pattern of discovery- and development-well land ownership as does the Sloss field. Apparently, the major firms had identified the areas in the vicinities of these fields as favorable for the occurrence of petroleum, but they farmed out or supported the efforts of the independent firms in drilling wildcat wells. In effect, the major firms used the independents to facilitate the evaluation of their acreage.

Results of development-well drilling and reserve ownership by class of firm for the entire basin are presented in Table 4.3. The major firms' development success ratio was substantially higher than that of any other class of firm. Further evidence that the major firms were able to retain higher quality acreage is

Table 4.3. Average Development Well Success and Volumes of Petroleum Reserves Credited by Operator Class from Fields Discovered from 1949 to 1979 in the Denver Basin

	Large majors	Large independents	Drilling contractors	Small independents	Total
Number of successful development wells	2326	1106	309	1926	5667
Number of unsuccessful development wells	1007	1342	533	2462	5344
Development-well success ratio	0.698	0.452	0.367	0.439	0.515
Reserves owned[a]	351.2	176.6	47.0	167.2	742.0
Reserves per development well drilled[b]	105.4	72.1	55.8	26.6	67.4

[a]In million barrels of producible oil.

[b]In thousand barrels of producible oil.

Source: Petroleum Information, Inc. (1975).

that their reserves developed per productive development well were 46 percent higher than those for the large independent firms, almost twice as high as those for the drilling contractors, and almost four times as high as the reserves per well developed by the small independent firms.

To me, it was important to be able to perform this analysis because it provided empirical data to support the qualitative statements that I had heard for years about how the major oil companies worked *with* the independents. The major firms did not do this out of altruism. I could now say that during the 1949–1974 period in the D-J Sandstone play, the major firms found 28 percent of the total petroleum discovered in their own wildcat wells, but their land ownership strategy allowed them to gain rights to 47.3 percent of the total reserves. Consequently, through their dealings with independent firms, the major firms' strategy of acquiring and holding large inventories of undeveloped acreage over the long term permitted them to gain production rights to oil and gas that they did not have to discover with their own wildcat wells.

The most difficult thing for me to explain was why, in the D-J Sandstone exploration play, the independent firms drilled so many poor-quality development acreage locations (Table 4.3). In their attempt to develop reserves, the large and the small independent firms and the drilling contractors drilled 4337 of the 5344 (81 percent) dry development wells. By definition, these wells were drilled in locations directly offsetting productive wells. Was it pure chance that caused them to be located just beyond the productive limits of the fields?

This behavior, particularly on the part of the smaller independent firms, is the result of the complex promotional devices used by these firms to finance exploration and development drilling. Typically, these firms have few fixed resources—small or nonexistent geologic staffs, few leases, and little, if any, drilling or production equipment. An important source of funds for these firms is money from individuals that would otherwise be paid as income tax. It is well known within the petroleum industry that many independent firms can and do make a profit even on a dry hole. When an independent firm promotes the funds to drill a hole, it usually does not consider the opportunity cost of using the funds in another endeavor. By contrast, the major firms finance drilling from retained earnings, and they evaluate the opportunity cost of using capital in a whole range of areas and prospects. It can then be said that the major firms' principal objective is to develop production for a profit, whereas the independent firms try to make a profit on the act of drilling as well as from production. Consequently, the independent firms are motivated to promote the drilling of as many holes as possible.

The explosive surge of wildcat and development drilling that occurs during the start of an exploration play and its falling discovery success ratio then can be linked, at least in part, to the ease with which the independent firms can promote the drilling of all types of wells. These independent firms promote speculative funds from tax shelter sources and obtain support from the major firms in terms of funds and acreage to drill wildcat and development wells. They are particularly successful when expectations are heightened during the early stages of an exploration play. As an exploration play matures, these expectations fall, and the ability of the independent firms to promote drilling wanes. To a considerable extent, the cyclical behavior of wildcat drilling during an exploration play is linked to the ability of the independent firms to promote the drilling of wells.

This concludes my discussion of the analysis of the behavior of exploration plays based on the empirical analysis of drilling and discovery rate data. For the most part, the rest of the chapter is devoted to a discussion of the work of Gordon Kaufman, who works at the Sloan School of Management at the Massachusetts Institute of Technology (MIT), and who, with the exception of Arps and Roberts, has made the most significant contribution to understanding the probabilistic (mathematical) structure that governs the behavior of exploration plays. Although Kaufman appreciates the contribution of empirical analysis, he does not really want to collect massive amounts of data, plot it on graph paper, push it through Fortran programs, and then attempt to extract meaning from wherever the investigation has taken him. Instead, he believes that the behavior of an exploration play is better determined by deducing its attributes from certain fundamental assumptions. This is the position that mathematicians commonly take as they approach their work, and it has an elegance that I admire. There is no argument about the practice of mathematics; it is either right or wrong, and the entire process can be checked exactly for errors. It is all very nice and tidy. There is no debate about sampling procedures, interpretation of data, or understanding an element of the science, which the empirical scientist has to deal with in nearly every aspect of his work. The mathematician may not even be concerned with whether his analyses are of value to any burning issue of the day. He is basically concerned with deducing the consequence of having made a set of assumptions. If the assumptions are well posed and the mathematics has been practiced without error, we will be presented with the truth when the mathematician finishes his work.

It was precisely from this position that Kaufman and his colleagues approached their treatment of the discovery process (Kaufman et al., 1975; Barouch and Kaufman, 1977). The concept of the exploration play was used as the basic underpinning for this analysis. It was assumed that the totality of all oil and gas fields occurring in a basin or region can be partitioned into homogeneous subpopulations. Each of these subpopulations would be recognized according to its distinct geologic characteristics. Each subpopulation of oil and gas fields is then confined to a petroleum zone, and the exploration play is the consequence of the wildcat drilling applied to the zone. Two critical assumptions are then made: *the size distribution of field sizes within a subpopulation is lognormal* and *the discoveries are made proportional to size and sampling without replacement.*

The manner in which these assumptions are stated predetermined how the analysis was to be done. Kaufman wanted to use only the above two assumptions. With them, he was able to position himself on a course for simplicity and elegance of exposition, analysis, presentation, and defense. A risk in taking this approach is that when its usefulness in application is examined and found to be flawed, the resulting attack is always against the assumptions. Kaufman was well aware of this pitfall and consequently put much effort into stitching together a basis for his assumptions. He appealed to his previous study (Kaufman, 1962), in which the adequacy of the lognormal distribution had been investigated. From these analyses, which were mainly based on log probability plots, he concluded, "The assumption of lognormality for the parent field size distribution was not unreasonable" (Kaufman et al., 1975, p. 118). He added further assurance by referring to the study performed on the size distribution of fields in Alberta by McCrossan (1969). It is my recollection that most resource analysts at the time believed that the lognormal was a good enough description for most applications. Kaufman even suggested that the law of mass action supported the assumption of the lognormality of the sizes of oil and gas fields. This assumption has been under attack recently because the results of empirical analyses pointed to the use of the J-shaped distribution (log-geometric) as a descriptor of the sizes of oil and gas fields and that the lognormal appearance of the observed size distribution of oil and gas fields is an artifact of the discovery process (Drew et al., 1982; Schuenemeyer and Drew, 1983; Attanasi and Drew, 1985).

Because of the prevailing wisdom in the oil industry, which stated that within an exploration play, the larger fields are discovered early and that the average size of the fields declines through time, Kaufman was able to justify his second assumption (Kaufman et al., 1975, p. 118). No data analysis was provided to support the assumption, although a vague inference in this regard was made by citing the analysis of Ryan (1973a).

In my conversation with Kaufman, he made it clear that, after stating these two assumptions and working away for awhile, he realized that he had charted a course right into the eye of a very difficult analytical problem. In his 1975 paper, he presented an apology for using a Monte Carlo procedure to determine the consequence of making these assumptions (Kaufman et al., 1975, p. 126) by saying that it gave an "intuitive feel" for the implications of the assumptions about the discovery process that he had made. This remark must be confusing to any reader who does not know that the Monte Carlo simulation procedure has always been judged by mathematicians to be an unsatisfactory substitute for a closed analytical solution. I have never been able to find out why mathematicians feel so strongly in this regard. It is not tht they do not trust a well-done simulation; rather, it is because of pride in their profession that they would prefer to have the problem left undone than to use the approximate solution provided by a Monte Carlo simulation. What Kaufman had found and did not say at the time was that the lognormal distribution was very difficult to work with when a sample was drawn proportional to size. So, in 1975, an "intuitive feel " was all that he would provide for the consequence of making these two assumptions.

Kaufman used the results of 60 simulation runs to construct the expected cumulative distribution graphs of the observed (discovered) field size distribution and the remaining-to-be-discovered distribution at various points in the evolution of an exploration play. The expected size of each discovery as a function of its position in the discovery sequence also was computed. An example of one of the expected cumulative distribution graphs is shown in Fig. 4.23. Here, pairs of graphs are displayed that represent the discovered (observed) and the remaining-to-be-discovered fields; for example, at the midpoint of this hypothetical exploration play of 100 fields, the cumulative distribution of expected discoveries is represented by the graph labeled "50" toward the upper left in the figure. The companion cumulative distribution of fields remaining to be discovered also is labeled "50" and is located toward the lower right of the figure. Note that the graphs of the cumulative number of discoveries bend upward in the right tail. This behavior has been observed in many empirical distributions of field size. When I was a graduate student, I plotted field size data that displayed this behavior. At that time, Professor Griffiths and I took it as evidence of a population of very large fields mixed with smaller fields. An important conclusion from reexamining these data was that they supported Kaufman's second assumption and are a mere consequence of sampling proportional to size.

The graphs produced from the simulations describing the relation between the expected size of discoveries and their order of discovery are displayed in Fig. 4.24. These graphs show, for example, that if the field size distribution is lognormal and if the discovery process can be described by sampling proportional to size, the first discovery made should be 12 times the mean size, given certain initial conditions. These conditions are that the exploration play contains a total of 100 fields, the population mean is 6.0, and the variance is 3.0

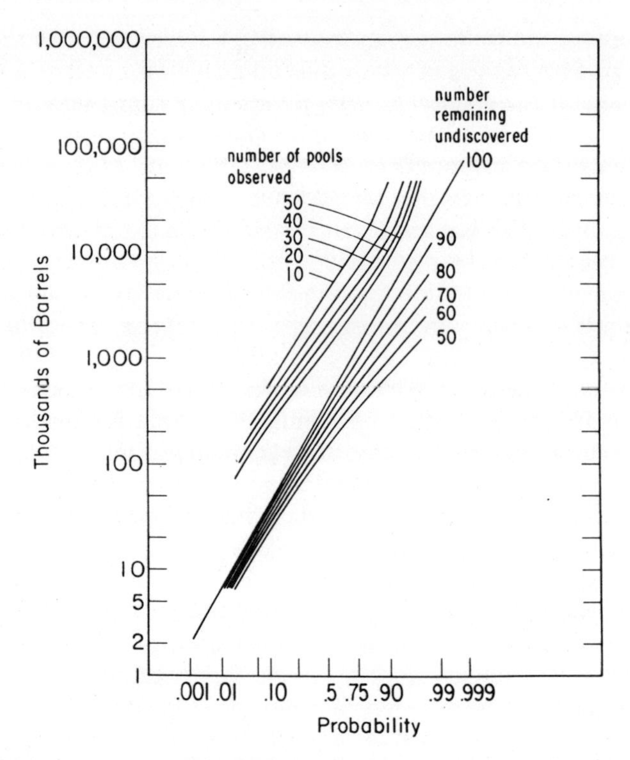

Figure 4.23. Simulated cumulative distribution functions for observed pool sizes and for pools remaining undiscovered when $N = 100$ (Kaufman et al., 1975).

(in thousands of barrels in the log space). When there are 1200 fields in the play, the expectation is that the first discovery will be 17 times the mean size. It is also to be expected that the mean size will fall very rapidly—so fast, in fact, that, if there are 100 fields in the play, only the first 20 discoveries are expected to be larger than the mean size of the total population. These results were the first of this type and stand today as a standard reference in the field.

Although this paper has been widely cited for its contribution, it and its companion pieces, which were published in the following years, were severely criticized. My colleague Jack Schuenemeyer and I were drawn into the fray by virtue of our affiliation with the USGS. The USGS has always had an image of extreme objectivity in its deliberation of issues in resource assessment. So it was natural that USGS employees would be called on from time to time, often in the hope that the image of our institution could be used to support a position held by the institution of the person extending the invitation. The most memorable encounter Jack and I experienced occurred in 1979 in the statistics department of Princeton University, where a group of professors had received a large grant from the U.S. Energy Information Agency (EIA). The purpose of this grant was to inspect the level and quality of the scientific effort in the field of oil- and gas-resource assessment. Gordon Kaufman's work had to be considered in any such evaluation. There was no doubt in my mind that the attitude at the EIA, and around the nation's capital in general, was that the topic had not been properly investigated because the best minds in the country had not yet grappled with the issues then existing in the field. It did not take great insight to read the tea leaves. All you had to do was go to a few meetings around town and watch and listen. I remember one meeting where I was told that I could only offer anecdotal information on the workings of the oil- and gas-exploration industry and that I had nothing to contribute because I had not used proper analysis techniques on a validated data set. I never did find out what constituted a valid set of data—only that it could not be collected by the oil industry. It was the temper of the times in the late 1970s in Washington, D.C., that the best analytical minds had to be brought into the field. The device used to bring these minds together was to involve the statistics departments of the major universities by offering them large research grants.

The statistics department at Princeton received one of these large grants to evaluate the quality of the research work in the field of oil- and gas-resource assessment and discovery process modeling. Many of the scientists who worked for industry or government in these fields were invited to come to Princeton to

Figure 4.24. Simulated means of size of *nth* pool discovered for finite population sizes N = 100, 150, 300, 600, and 1200 (Kaufman et al., 1975).

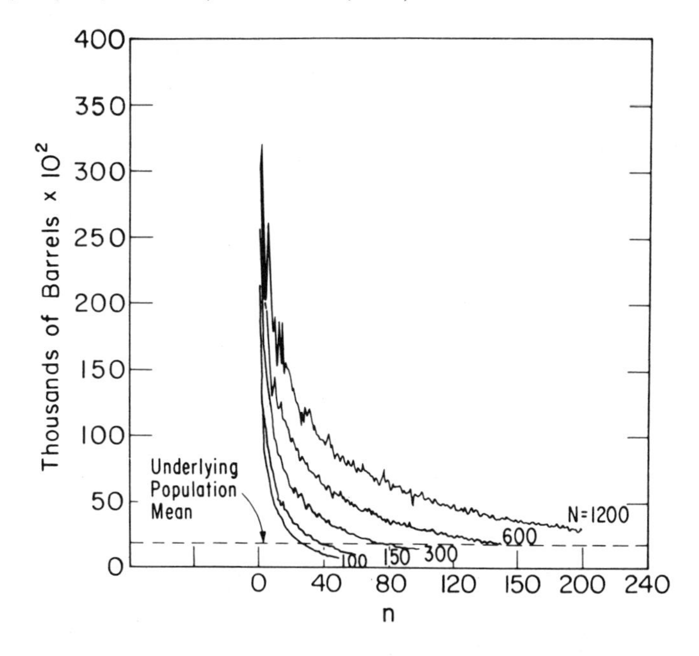

discuss the value of the analytical procedures in use and possible improvements that could be made. Jack Schuenemeyer and I were invited for the expressed purpose of talking about our research work in the field of discovery process modeling.

We had no sooner arrived when we were challenged to speak on the validity of sampling proportional to size argument that Gordon Kaufman had invented. Before we could speak, we were told that it was clear from their analysis of the history of the discovery of the oil and gas fields in the state of Kansas between 1900 and 1975 that Kaufman's assumption was wrong (Bloomfield et al., 1979). They had concluded that the discoverability coefficient in his probability of discovery equation (4.1 below) was much smaller than the value of $\alpha = 1.0$ that Kaufman had assumed.

$$P\{(1,2,\ldots,N)\,|\,A,\alpha\} = \prod_{j=1}^{N} A_j^{\alpha}/(A_j^{\alpha} + \cdots + A_N^{\alpha}) \qquad (4.1)$$

They had concluded from their analysis of data from Kansas that $\alpha = 0.33$.

We tried to explain that Kaufman had intended this assumption to be used only in an individual exploration play. I remember arguing that at least eight plays had unfolded in Kansas during the 1900–1975 time period, each having a quite different set of characteristics. We put forward the idea that even given this assumption's validity at the level of each individual play, no one could say before the fact of analyzing the data what the aggregate profiles would look like. It was obvious to us that this was not the kind of information that these professors of statistics had desired. We had brought a millstone to hang around their necks, and this was not an acceptable gift. I tried to soften the blow by suggesting that they call on Bob Atkin at the Department of Energy who really knew the history of the petroleum exploration business in Kansas; he had overseen the drilling of more than 100 wildcat wells over a 21-year period and could identify the exploration plays. I even called him later and told him what had been recommended. He groaned and said he would do it. However, he was never asked to work on the data. Some time afterward, I concluded that because such an effort in data reduction was too complicated and too labor intensive, it was the wrong kind of work for the academic world.

It was soon obvious in our meeting at Princeton that we were wasting our time. After Professor Geoffrey S. Watson said pointedly to me that he could deduce everything that I concluded from the analysis of data, I remember dropping the chalk in the chalk tray, stepping away from the blackboard, and saying to myself that it was time to leave. We had delivered our news, and it had been rejected.

Going home on the train the next day, we wondered what was to come next. We knew that this was a premiere study and that a lot of folks were watching for the results. To us, it was not so much that their minds were made up, but that they were reluctant to take a wholesome attitude toward the idea that the problem at hand was essentially a problem to be solved by data analysis. Their approach to data analysis had been too simplistic, and that was that.

After the Princeton report was finished and circulated to the public, the next move was up to Kaufman. His response was predictable. I knew it was not going to matter much from where he chose to start his rebuttal. It was going to be like one of those football games where the veteran quarterback picks the defensive secondary of the inferior opponent to pieces by using a deadly aerial attack. He made all the points that Schuenemeyer and I had made. He did, however, keep it all gentlemanly (Kaufman and Wang, 1980; Kaufman, 1983, pp. 249–251). Kaufman went on to show that the maximum likelihood estimator used by Bloomfield et al. (1979) was not robust enough for even moderate sample sizes (Kaufman, 1983, p. 258–261). At the USGS, we saw the controversy as being pretty academic and went on with our analyses of discovery process models and field size distributions.

5

The Economists

In the previous chapter, a discussion of the motives of the different types of exploration firms within the context of exploration behavior was presented without using the formal vocabulary of microeconomics. In this chapter, my interaction with economists on oil- and gas-resource assessment issues is told. This is a story of conflict between two academic disciplines that are based on very different analytical structures. As I became involved with economists in discussions on the future supply of oil and gas during the middle 1970s, I found that the only way I could hold my own was to study the field formally.

At this time, not only was the future supply of oil and gas an issue for many analysts in the nation's capital, but so was the supply of most naturally occurring raw materials. As I was sent to meetings around town where these issues were being discussed, I became aware that the dominant intellectual position was that of the economist. I cannot remember an incident when these folks were not very sure of themselves and quick to belittle any conclusion reached by any means other than the use of economic and econometric analysis. The earth scientists who interacted with these economists were invariably confused by this attitude. Although there were countless incidents that I could mention to illustrate this conflict, I have chosen four interchanges that I believe were typical.

I was able to gather one by watching Allen Clark, the fellow I worked for, struggle with a well-trained economist employed by our mother agency, the Department of Interior (DOI), over an issue that involved the location of the major mines of the world and what they produced. Clark had the traditional mindset of the exploration geologist, and it did not serve him well this day. He knew the nuts and bolts of the world mining scene but had not yet become familiar with the calculus of economic reasoning as applied to world trade. He did, however, have the zeal to learn. He even went so far as to purchase a copy of his adversary's Ph.D thesis, which was on a topic in international trade, from University Microfilms Inc. Some time later in his office, I picked up the blue-bound thesis and asked how it went with the DOI economist. His response was a shrug. I noticed that the thesis was page after page of partial differentials. Clark said that he never was able to convince the economist that raw material supply was a topic worth studying because the economist argued that there would always be an adequate supply of these materials. They were to be taken as a given in all economic analyses.

The second experience also was collected at the expense of the tribulations of a coworker. This gentleman was Don Brobst, who was the USGS expert on barite, a mineral commodity used in oil well drilling muds. If the truth be known, Brobst may have been the world expert on this commodity. He had seen every mine site that produced this commodity in the United States and most of the places where this commodity was mined in Canada. He had read extensively about every major occurrence of this commodity in the world. He knew its geology, mineralogy, geochemistry, the mining methods used, the grades, the tonnages of reserves, and even a lot about the cost of mining. He was an important somebody to know with regard to this commodity. If you had a question about this commodity, you went to him for the answer.

Brobst found that his knowledge and experience was of little use when he attended a high-level, week-long meeting on the issues of raw material supply at the Brookings Institution in downtown Washington, D.C., in 1976. He was told by an eminent professor of economics that he had nothing to contribute to the subject of resource economics and, what was more, that the field of geology had nothing to contribute to this subject either. I remember that when he returned to work he was irate. He would ask over and over again, "Who was this economist to fold his hands across his chest and utter such words to me?"

This interchange was most confusing to me. I could not understand it at all because if you wanted to discover a new deposit of barite, you would go to Brobst for advice and council. This fact meant absolutely nothing to the economist when it came to the issue of resource economics. The economist was a bright, well-read, and knowledgeable man. He had to respect such an ability to contribute to the general welfare. What made him say the things he said? On the surface, these two men would say they were talking about the same thing when they used the phrase "resource economics." There could not be much common ground because they talked *at* each other, and little or no meaning was conveyed.

The third incident I experienced firsthand when I was sent to a meeting to listen to the preliminary results of a comprehensive effort to forecast the future supply of natural gas. The audience was a "Who's Who" of energy economics for the DOI and several other federal government departments. At this time, the Department of Energy (DOE) had not yet been created, so DOI still had a major position in the field. The principal investigator was Paul MacAvoy, professor of economics at MIT. He made a strong and certain delivery. The punch line was that the natural gas market would clear at a selling price of 50 cents per thousand cubic feet at the wellhead. At this time, the average wellhead price was around 20 cents per million cubic feet, and there was a shortage at this price. So, 50 cents would do the trick and clear the market. A market-clearing price is one at which suppliers are willing to supply exactly what consumers are willing to take. MacAvoy presented viewgraphs of price and quantity relations. I was struck by thc lack of scales on the axes. Many regression equations were presented; there were several thousand equations and variables in the econometric model that he and his coworkers had built. I started to lean forward in my chair to ask a question about the gas discovery module. Realizing that I would be struck down by a mighty blast from MacAvoy, I sat back to collect more data. Toward the end of the meeting, an offer was made to discuss the specification of the model. MacAvoy said that they did not know everything about the natural gas business, and, if any of us would like to help them make the model better, he would listen. I was pretty sure this was a throwaway phrase that was tossed in at the end to round out his presentation.

To a physical scientist like myself, this was a new world. Here, equations were used to estimate the size of new discoveries, and the price variables were mixed right in with index and dummy variables. I knew that a dummy time variable was used to capture a systematic effect in a data set not picked up by the other independent variables, but this stuff was a long way beyond that. Another fact I collected was that most of the price effect estimated in the time series came out of cross-sectional differences.

Several months later, at a workshop on energy modeling held at MIT, this econometric model and its conclusions were presented again. The junior author of the model, Bob Pindyke, made the presentation this time. I could not be quiet any longer. I rose to argue the point that the price effect estimated out of the cross section was most likely the result of differences in hydrogen sulfide and carbon dioxide contents of the gas and crude oils rather than anything specified in the model. I wondered openly about using cross-sectional differences to estimate time-series effects. The rebuttal did not come from the speaker but rather from the chief economist of a major oil company, who, like myself, was a member of the audience. His counterpoint was to the effect that what had been done had been done all right. At the break, I asked the chief economist why he said such a thing. Was it not true that what had been done was little better than finding a correlation between the rainfall in California one year and the number of babies born in New York City the next year. He could not defend the specification of the equations. He was able to say with confidence only that that was the way they do it in econometrics. His real point was that I was not an econometrician and what did I know anyway.

The fourth story I will use here is presented to expand the image of the interaction between the government's geologists and economists on the issue of resource economics. I participated in this fourth experience firsthand as a reviewer of a National Science Foundation (NSF) proposal for designing a national energy accounting system. The program manager from the NSF had expected a calm meeting at which support for the proposal would be collected. Unfortunately for him, the proposal showed many signs of poor construction and of being written in a hurry. Before the afternoon was over, the proposal, its author, and the vice-president of the consulting firm who had come along to watch would be badly beaten. The program manager from the NSF would wonder for some time just what had happened. As I remember it, the technical discussion quickly turned against the idea of a national energy account being kept on the basis of British thermal units (Btus). The ebb of the meeting came when Walt Dupree, an energy analyst from the U.S. Bureau of Mines, gave what I later came to call the energy-accounting equivalent of the Cross of Gold speech. This analyst rolled up the proposal in his hand, pointed it at the economist from the consulting firm, and bellowed, "The currency of energy is not Btus!" He struck at the heart of the proposal saying, "A Btu of coal is not a Btu of oil." The vice-president from the consulting firm wrote notes furiously to himself and cast cold hard stares at the earth scientists and engineers who had been invited by the NSF to review the proposal.

One of the worst statements in the proposal was that oil and gas are produced from the Earth's soil. I informed the men from the consulting firm that soil was the top few inches to at most several feet of the Earth. My remark ignited their tempers. I could have said, had I been quick enough, "The only oil in the Earth's soil has dripped out of the crankcase of the tractor that plowed it." The meeting was terminated at about this point because it was obvious that no useful purpose was being served.

Because of such experiences as these, I came to wonder why the economists allowed themselves to have such strong opinions on a field of investigation in which they had little or no interest in learning the technical aspects. It seemed that they did not even care to learn the vocabulary of the field. They had their own vocabulary, and it became obvious to me that before I was allowed to contribute to any conversation, I would have to learn it.

My colleagues at the USGS found my interest in trying to deal with economists by learning economics to be most unusual. King Hubbert was sure I had taken leave of my senses. Better I take a course in Laplace transforms and learn something useful. However, I had made up my mind to go formal and learn the stuff. The department chairman at a nearby university who took me in was trained at the University of Chicago. He was amused by my reason for knocking on his door. I told him that there were a lot of terms that had gone by me in meetings, discussions, and phone calls. I told him that, at minimum, I absolutely had to learn the vocabulary used by the economists. Some of the words were in the vernacular, but I knew we did not use them as an economist meant them to be used. These were words like "competitive," "profit," "rent," "substitute," "long run," and "short run." There were other words and phrases that I had to learn that were not used in the vernacular, such as "elasticity," "normal good," "income effect," "multiplier," "production function," and "endogenous." My real quest was to find out why the study of the technical issues in the field of natural resource supply, which I valued so highly, was insignificant to the economist in his study of resource economics. I would be on the right track when I learned about production functions and factor inputs. The lock would not open, however, until I used the right key, which I would get after attending lectures on substitutes and elasticities.

The department chairman chuckled as he listened to me describe my desire to gain a facility in economics. He reminded me of my family doctor, reassuring me that my aches and pains were only part of being full grown, as he said, "Sure, we can fix you up." The chairman smiled as he wrote a prescription for course work for the poor troubled earth scientist who sat before him. There would be three courses in microeconomics—price theory, as he called it. He suggested three courses in macroeconomics, four in mathematical economics and econometrics, and four or five electives. He suggested industrial organization, social welfare, and international economics. I left his office glad to be finished with the interview and looking forward to my journey toward understanding what makes economists tick.

I went to class wondering if the instructors would reveal what I was seeking. In time, they did, but I had to stitch most of the ideas together myself. One of the first hints came from a lecture in a mathematical economics course, when the instructor talked about making a decision to consume children. He said a child, which is a consumer-durable product that creates an implicit income stream, is to be valued in a trade-off against other consumer durables, such as houses, cars, and refrigerators. What a shocking statement! I never heard of valuing a child in terms of the usefulness of a refrigerator. What is the stuff of economics if they talk like this? I was to find out that, to the economist, every item imaginable is equatable to an income stream that can be valued in terms of its utility against the utility of all other income streams. Some income streams are tangible, such as monthly interest checks from a savings account, and others are intangible, such as the joy one gets over the years from owning a painting or a gold watch. The argument would conclude that, whatever you add to your consumption bundle, it is done by making decisions about the utility of income streams.

The instructor went on to say that the important thing for us as economists was that we could explain why households are consuming fewer children these days. It has to do with the rising wage rate for women. He argued that more and more women were trading off the income stream generated by having another child for the income stream generated from the joy of work. So, here I had my first clue about the thought processes used by economists. They differ from geologists because they are interested in examining the motives of people, and they tend to look on physical things as being nonspecific. In contrast, geologists are interested in the history of the inanimate Earth and have a highly cultivated sense of the specificity of physical things; for example, geologists give the names of places, people, and even feelings to every mine and oil field, thereby elevating the inanimate to the most specific of categories.

The geologist has the propensity to identify everything as being unique. No detail is insignificant. Geologists, then, are reductionists. Their motto might well be, "Everything that can be perceived is different, and if you don't see a difference, keep going in your examination until you find a difference. Then lump things back together only if you have to." A counter trend of a sort has been established lately in geology where the ideas of plate tectonics have placed the emphasis on the study of large-scale phenomena.

The economist, however, would be perfectly comfortable with oil fields and mines being identified by numbers and only vaguely defined in a physical sense. Frankly, the economist is confused by the geologists' desire to specifiy uniqueness through labeling. The economist sees the purpose of human intellectual activity as understanding the manner in which choices are made. He is particularly interested in the adjustment of choices when prices change. The economist sees these choices as transactions being made at an intangible place that he usually refers to as "the margin."

The best lesson I had in economics came by putting together several ideas from lectures in three different economics courses. In a macroeconomics course, a fellow student asked the professor what he thought about the long lines of cars waiting for gasoline during the oil embargo. The professor responded that he knew that there was nothing to worry about because the lines were a short-run adjustment on the demand side and also that the Organization of Petroleum Exporting Countries (OPEC) could not keep the price up because the market structure for crude oil must be competitive. He finished with the idea that there could not be much of a barrier to entry into the crude oil business. This was the first part of my lesson.

In a lecture in a microeconomics course taken at about the same time, the professor directed a pointed remark at me when I tried to argue that there was something special about natural gas. He said, "Sweaters are a substitute for natural gas! It is just a factor input. Oil, gas, and solar, they are all fungible." I could imagine millions of houses across the country freezing solid. It would be like the end of the world. The professor did not see that at all. He said that it would, of course, be messy for a short time but that we would adjust quickly. This was the second part of my lesson.

The third bit of information came when the professor of a course in industrial organization made a solid case that the airline industry was a competitive industry. The barriers to entry to this business had all been put in place by regulation. One of my fellow students mentioned that it costs hundreds of millions of dollars to get into the airline business. The professor said that this was not a barrier to entry; $100 million was nothing. He went on to say that there were many people who can fly planes; there was nothing special about being a pilot. There were also lots of routes to fly. A little bit of credit and a phone call would get you a 727. There was nothing unique anywhere in the business. It is like growing corn or wheat; lots of people can do it. There was no rent to be captured except by erecting an artificial barrier to entry through regulation. Another student remarked that a lot of airlines could go out of business. The professor replied that that could happen but a lot could also enter.

So there it was. I had the rudiments of an answer to my question. The escape hatch from how a geologist thinks to how an economist thinks was opened when I linked together the following string of statements and images: The oil and gas business is not unlike the airline business. There is nothing special about oil and gas. It is just a factor input to a production function. The uniqueness that I, as a geologist, ascribe to these commodities is the result of my propensity to know through detail labeling. There is nothing romantic about a barrel of oil. One barrel is just like another. Indonesian crude burns just as well as west Texas crude. The same dollars will buy either in the long run. If the price of oil and gas goes up, substitution will occur. Remember, everything has substitutes! Above all else, remember that. Not using oil is a substitute for not having it. Sweaters are a substitute for it in the winter, and sweating is a substitute for it in the summer. Smaller cars are a substitute for it. Wood is a substitute for it. And last, but by no means least, the sun is a substitute for it. Also, on the other side, you must always remember that when the price goes up, there are people who will jump in and start producing oil and gas. Do not worry about who these people are; trust me, they are there. The action of these suppliers will build a force against the rising price. Our economic lives are all linked together, and we can live with less oil and gas at higher prices. If the price goes high enough and we have a little time to adjust, we will find complete substitutes for oil and gas. You can go look at it in a jar in a museum case if the price goes high enough, and life will go on. This may be the ultimate definition of not being unique. Economists only care about the supply of crude oil as an energy source, and there are lots of energy sources.

Well, what was I to do with my new-found insights? One of the first things I did was think about the conflicts that must exist within OPEC as it tried to maintain power over the price of crude oil by controlling production. I imagined arguments between those who, like certain Iranian officials, wanted the price pegged at $100 per barrel and those who knew their price theory and were trying to find a price at which consumers of crude oil would not start the process of substituting another energy source as a factor input and keeping new suppliers from appearing on the world market whose production would start to drive the price back down. They knew that, at $100 per barrel, some crude oil would be sold but that the elasticity of demand would decrease their total income because, at this price, less crude oil would be consumed and massive substitutions would occur. The net effect of a $100-per-barrel price would be technological development in the solar energy and coal conversion processes. It is also the business of the economist to assess the motives of each side of the debate and to tie these motives to the economic character of the country they represent.

On the more personal side, I had an experience with a colleague that sums up what it is like to cross the bridge from geology to economics and then to look back across to the other side. It occurred while I was returning from a meeting in downtown Washington, D.C., where a colleague had presented the USGS's new assessment of U.S. oil and gas resources to a group from the Exxon Company. My colleague, a geologist, looked out of the car window as we drove through mile after mile of the mansions of Great Falls and McLean, Virginia, along Route 193. He said, "It's all doomed. This way of life is doomed." I knew why he said this. He saw the U.S. resources of conventional crude oil and natural gas as being small and dwindling fast. He could not imagine anything but massive change coming in the near future. Maybe only a wasteland would be left where these mansions stood. It is easy for the geologist to imagine that we would all have to move into crowded quarters in the winter just to survive on the little heat that would be available.

Now, if I had argued as an economist, I would have made the following points. From the demand side, we start by noticing that these households are wealthy. Wealth means they have options, and, because they are human, they can and will adjust. Technically, they will substitute. A massive rise in the price of energy will be a signal to them. They will react by closing off two or three rooms in their houses. They will reduce from four cars to three cars, maybe even cut down to two cars. They will buy smaller cars. They will buy diesel cars. They can pay $5.00 or $10.00 for a gallon of gasoline. They will buy solar hot-water heaters. They will forego a trip to Europe if the price of gasoline goes to $15.00 per gallon. You see, it is all very fluid inside of the consumption bundle of the wealthy.

At the same time that this response is occurring on the demand side, there is a companion response occurring on the supply side. It is dogma in economic theory that the price rise that is a signal to the households in McLean and Great Falls to begin conservation and substitution is also a signal to suppliers to increase the supply. Had I said this to my colleague, he would have turned to me and said, "Just where are those new supplies going to come from? Point to the spot on the map where they are going to be found! In what basins and in what rocks are we going to find these new supplies? We have found the lion's share of the conventional oil and gas around the world. There is still a lot to be found but not in comparison to what we have as reserves today. We have probably used more oil and gas already than there is left to be found in the conventional type of oil and gas fields remaining to be discovered. Sure, there is a lot of nonconventional stuff around in the form of shale oil and tar sands, but these sources cannot be compared to conventional crude oil flowing to a well bore! You have to mine that stuff and cook it and recycle the waste. It is just not crude oil as we know it."

In general, it can be said that most, if not all, geologists cannot see a smooth transition from conventional supplies of crude oil to nonconventional sources. Conventional crude oil stands alone by itself; it is unique. It is unique because it flows to a well bore. It is unique because we know so much about the geography and geology and where it comes from. It is unique because we know many places where it does not or cannot occur. This hard negative information may be the most important information that the geologist has to form opinions about the future of potential supplies of conventional oil and gas.

My colleague is correctly exercising his right as a scientist when he demands that the economist show him a physical place where oil and gas is going to be discovered when the price rises. The economist believes that it is going to happen because he knows the motives of humans, and he has a large body of history that suggests that his economic theory always has made useful models. The economist has watched many doomsayers be bowled over by human adaptation to changing circumstances. As yet, the geologist has not been able to penetrate this mindset using arguments that make the economist stop and think twice about the "resource economics question." In a nutshell, the economist cannot believe the idea that, for the first time, we are standing before an insoluble problem with the future supply of natural resources. To him, it is all a matter of substitutes and what an economy will trade off as the prices of things change with time.

So, I see the conflict between the economist and the geologist occurring across a gap that will probably not close soon. I see the geologist who holds tightly to his view of a finite Earth and its dwindling conventional resources versus the economist who sees nothing to worry about as he looks back over our history. An individual economist might go so far as to say, "Conventional resources of oil and gas are limited, but a rising price will send a signal to all concerned, they will respond, and the markets will clear." As soon as the geologist tells the economist that there is a vast, nearly infinite supply of energy in the nonconventional sources, the economist will smile securely and begin to think about how the economy will adjust to obtain energy from these sources and what other changes will ripple through it in the process. The geologist will raise his hand, warning the economist again that he does not know what he is getting into because there are orders of magnitude of change to be dealt with. The economist will hear the geologist but will not listen.

On that day when my colleague and I drove through the opulent Virginia countryside, I chose not to argue as the economist. My colleague had a temper and was easily excited. Besides, he had a point that cannot be overlooked by the economist. It is possible that the supply of crude oil could be interrupted very severely, and contributing factors, such as a cold winter, could also occur to spell disaster for us all. The probability of such events occurring together is very small but no so small as to be disregarded. The United States has built, at great expense, a national petroleum stockpile to guard against just such a set of events.

As I watch the economic transactions going on today at the boundary between conventional and nonconventional petroleum supplies, the biggest unknown to me is what a major transition in volume across this boundary will look like in a quantitative sense. We still live in a world of conventional supply, and it is difficult to look into the future and to separate sense from nonsense in the available promotional literature and information. There are plants that convert coal to liquid fuels in South Africa, but only the South Africans know the full cost. The major oil companies have mostly pulled out of the development of oil shale.

I suspect that the economists are going to have the last laugh on those of us who prefer to analyze the future supply of energy commodities measured in barrels and tons accompanied by long lists of facts about where these commodities occur. This approach yields an intuitive feel for our situation, but this is no match for the supply-and-demand calculus that the economist has in his tool chest. I suspect that it will be to the delight of the economists that analysts such as myself will be frustrated, for example, by the comingling of oil and gas from all future sources. If and when shale oil comes on-stream, it is going to go into the same pipeline as the conventional crudes and come out the other end of some refinery as 89 octane gasoline. If a question is asked about what effect a specific economic stimulus had on aggregate supply, the economist will stand in front of the blackboard and draw a supply-and-demand curve and state his conclusions. We earth scientists might want to answer the question in another manner or perhaps send a warning that based on what we have found from our scientific investigations, future supply might change. We are, however, at a disadvantage because, by the nature of our subject matter, we cannot make the tidy summaries that appear to be ladened with the information that the economist provides when he draws two lines on the blackboard and labels one supply and the other demand.

6

Discovery Rate Forecasting, Part 1: The Permian Basin

This chapter describes the application of discovery process modeling to the Permian basin for the purpose of forecasting the future rates of oil and gas discovery in this very large region of the United States. This project was initiated in 1976 to fulfill one of the requirements specified by an interagency task force commissioned to estimate how much oil and gas might be available in the future at various prices from all potential sources in the United States. Although the USGS was designated as the lead agency, a large part of the total effort to complete the report was made by personnel in several other U.S. government bureaus. Before describing the details of how the forecasting of future discovery rates in the Permian basin was accomplished, we will go back in time and set the historical stage that led up to the commissioning of the interagency study.

Perhaps the easiest way to do this is to combine my own recollections with material contained in the two excellent reviews on oil and gas discovery rate forecasting by Gordon Kaufman (1983) and Deverle Harris (1984). One does not have to read very far in these two books before an important fact is uncovered. In the not-too-distant past of the 1960s and early 1970s, there were two opposing points of view about the size of the undiscovered inventory of oil and gas resources of the United States. On one side were the large estimates made by Vince McKelvey and his colleagues, who forecast a 500-billion- to 600-billion-barrel estimate for the ultimate productivity of conventional crude oil in the United States. King Hubbert was the opposition with his estimate of 170 billion barrels. To me, it was pure coincidence that both these views were harbored within the USGS at the same time.

By the time I came to the USGS, the debate had been going on for so long that both sides were entrenched. I quickly found myself sitting on the fence by saying that Vince McKelvey's estimate was perhaps too large and that King Hubbert's was maybe too small. Part of the problem was that I never could determine exactly which categories of crude oil were or were not to be included in the McKelvey and Hubbert estimates. As the new kid on the block, the fence had a certain appeal.

In 1972, nobody could have foreseen that, in only three years' time, the USGS would change its assessment method from one based on a volumetric yield analysis (the McKelvey method) to one based on the analysis of petroleum geologic data using expert judgment. This assessment was published in USGS Circular 725 (Miller et al., 1975). By using this newly implemented methodology, an estimate was produced that was very close to the estimate made by Hubbert in 1972. Gordon Kaufman (1983, pp. 113–125, 148–163) included an excellent treatment of this adjustment in his book. In general, it was believed by the crowd of concerned onlookers that Circular 725 brought a peaceful solution—the official USGS position had been brought into line with the reasoning of oil industry analysts and, of course, King Hubbert. For those readers who wish to probe this issue further, I recommend reading Harris (1984, pp. 349–354), who presented evidence contrary to the idea that a peaceful settlement was achieved with this circular and that everybody was in agreement. In this analysis, Harris gave a didactic tutorial using probability theory. Vince McKelvey was Director of the USGS at the time this circular was published, and it is my recollection that, while he did not fully believe in either the techniques used or the results, he worked hard to effect a smooth transition from the older, larger oil- and gas-resource estimates to the newer, smaller ones.

In passing, it is worth mentioning that the strongest feelings about the downward revision presented in Circular 725 were expressed to me by Bill Vogley, a ranking economist in the DOI. He said that the USGS had capitulated to the pressure of the environmentalists. I wish he had written his position down so that I could reference it and explore the reasons for his statement. The connection to the environmental issue, at least at that time, is through an argument that goes something like, "If there is little or no oil and gas left in a region, then there is nothing much lost to mankind by locking out the oil industry and keeping the place safe for beast, fish, and fowl." I guess I was supposed to have concluded that large resource estimates would have kept the environmentalists at bay. I thought this argument was an odd contraption then and have not changed my opinion today.

The problem that Vince McKelvey had with the methods delineated in Circular 725 centered on the fact that some use of the discovery rate decline techniques had been made in producing these estimates. It will be of interest to the historian to see whether McKelvey or Hubbert has the last word in their debate on the usefulness of decline extrapolation techniques. Hubbert was a believer in the validity of their use, and Vince McKelvey was not. As best as I can reckon, McKelvey currently has the last word by virtue of his remarks on page 343 of a paper presented at the International Geological Congress in Moscow (McKelvey and Masters, 1984). In this paper, he stated his opinion of decline extrapolations: "They do not provide a means for estimating resources of unexplored regions; the human activities they analyze are strongly influenced by economic, political, and technologic factors that are at best only indirect indications of the amount of oil and gas in the ground; and there is no reason to believe that the patterns of the past accurately foretell the future (Ryan, 1965; Harris, 1977)."

The operative words in the above summation are "at best only indirect indications of the amount of oil and gas in the ground." Vince probably overstated his case a bit because decline extrapolations never have been made by a serious analyst on a region where drilling and discovery data are sparse. Here again, we have the same sort of philosophical disagreement that I said existed between the geologist and the economist over the resource economics issue in Chapter 5.

The nub of Vince's position was that he simply refused to believe that a time series of numbers said anything real about the undiscovered resources of an unexplored (poorly explored, I think he meant to say) region. Vince believed that the real meaning of resource assessment is contained in geologic concepts and that little meaning is contained in past discovery and production rate data; it is more meaningful to put your trust in tangible geologic observations than in something that has a large human element in it, such as the consequence of a decision to drill a well. To Vince, dealing with the outcome of decisions meant dealing with the intangible. Vince believed that it was safer to stick with the rocks. If put to a vote at, say, the annual American Association of Petroleum Geologists (AAPG) convention, Vince's point of view would be well taken and could easily carry the day. Numbers are numbers, and they have their place, but geologic principles and data are the reality of resource assessment.

To those of us who took up the discovery process modeling part of the resource-assessment business some years ago, the issue of whether Vince McKelvey or King Hubbert had made the better estimate of the ultimate productivity of conventional crude oil and natural gas in the United States was only a tangential issue. As a group, we were interested in the mechanics of how undiscovered resources are converted to reserves through the exploration process. We also were concerned with the task of finding out which statistical distribution best described the sizes of oil and gas fields. It may seem impossible to believe today, but, during the early 1970s, we were, in fact, investigating the idea of whether and to what degree larger oil and gas fields are discovered earlier than smaller oil and gas fields. We wanted to know how regular this phenomenon was across exploration plays, basins, and countries.

It was also important for us to see if price and cost changes affected the rates of discovery or if the discovery process was, as many believed at that time, like a juggernaut that pushed its way toward completion without much regard for external control. I knew quite a few people who thought that finding oil and gas was an adventure in a world of risk and opportunity in which the cost of a foot of drilling or the price of a barrel of oil was of little concern. A sage in the business once summed up the whole activity of exploring for oil and gas by saying, "Oil must be sought first of all in our minds" (Pratt, 1942). With this remark, he was doing more than telling oilmen to think carefully about the evaluation of their exploration prospects to reduce the risk of failure. It is my belief that Pratt was emphasizing the frontier-type adventure that awaits the oilman as he plots the course toward the drilling of a wildcat well to test his ideas of where oil and gas might be found.

By the nature of the problems on which we discovery process modelers had chosen to work, we were forced to focus our attention on small-scale studies, much smaller than the issues being debated in the assessment of the ultimate productivity of the United States as a whole. As was pointed out in Chapter 4, we worked, for the most part, at the level of the exploration play. It would depend on which geologist you asked, but most would say that 200–300 exploration plays have unfolded in the United States, thereby making analyses at the exploration play level very small scale in comparison to a national assessment. It may be important to note that even though we call exploration plays small-scale phenomena, they can involve large areas of real estate, not uncommonly 10,000–20,000 square miles each and in which as many as 30,000–50,000 wildcat wells were drilled. The contribution that discovery process modeling could make to the assessment of the undiscovered oil and gas resources of the nation was soon recognized.

With the publication of Circular 725, the seed was sown for the connection between the discovery process modelers and the geologists at the USGS who were making and defending their large-scale oil and gas assessments for the United States. It has been one of our best-kept secrets that a small, but important, section of text was intruded forcibly (geologic pun intended) into the first page of the circular by none other than a highly placed economist who worked in the DOI. His name was Hermann Enzer.

I knew the hearsay about the origin of this section of text and have made use of it in debates over the years as an example of how emotionally charged the air over the oil- and gas-resource assessment issue was at the time of the publication of this circular. It was at the height of the energy crisis, and the adequacy of the future supply of oil and gas was uncertain and debated vigorously and emotionally. Before I could feel comfortable using this story here, I had to know for sure whether the hearsay was true. So, not long ago, when the opportunity arose, I pointedly questioned Enzer as to whether he was the author of the text, and, as best as I can tell, he admitted to writing these words and that his boss, Jack Carlson, assistant secretary of the DOI, "talked" the Survey into placing it in the second column of the first page of the circular. Knowing the origin of this statement is most important because, to my knowledge, it is the first time in a USGS publication that the size of the undiscovered recoverable resources of a commodity was specifically related to the price of the commodity and the cost of discovering and extracting the commodity. Of particular significance was the statement that the size of the undiscovered resources would increase by a given percentage as the price continued to rise over the cost. The exact text that is relevant here is, "The higher price-cost ratios existing in 1975, if they should continue or increase even higher, would likely increase estimates of both undiscovered recoverable resources and reserves significantly—some economists think perhaps by half again. This possible added potential is being considered in a follow-on study planned for completion within a year." The last sentence of this text became the franchise for the discovery process modelers, the petroleum engineers, and the economists to formally enter into the field of large-scale petroleum-resource assessments at the USGS. The stage had been set by circumstance, and all we had to do was show up and play our parts.

In August 1976, Richard Meyer was selected by the chief geologist of the USGS to be the program coordinator of the follow-up study promised on the first page of Circular 725. To manage this complex assignment, Meyer formed a committee to decide what work should be done. This committee limited the study to three regions: a mature producing area (the Permian basin), a partially explored area (the Gulf of Mexico offshore area), and a frontier area (the Baltimore Canyon basin). His goal was to produce a whole new type of oil- and gas-resource assessment in which the disciplines of geology, petroleum engineering, and economics would meet and contribute to the production of a balanced product. Meyer knew he had taken a risk in that he might not be able to direct the diverse cast of characters that he had assembled. Meyer ensured that we all knew what we had agreed to do by preparing and having the committee members sign an interagency agreement involving the DOI, the USGS, the U.S. Bureau of Mines, the Federal Energy Adminstration, the Federal Power Commission, and the U.S. Energy Research and Development Administration (now DOE).

During one of the early meetings called by Meyer, the question of how the discovery process modelers were going to produce the exploration function was asked; this was the task that I had been assigned to lead. It was generally believed by the members of the group that this task was exceedingly difficult to carry out. Much of the problem that I faced was convincing the group that discovery process modeling could be used effectively for this purpose. This was to be expected because little was known about these models. At the USGS, Jack Schuenemeyer, Dave Root, and I had developed a discovery process model based on the physical exhaustion principle for use at the level of the exploration play. Because we had not yet published any of the results, we had to provide more than the normal explanation for how we intended to proceed. In addition, the Arps and Roberts discovery process model that had been developed nearly 20 years before was virtually unknown. I had read the description of this model, but, as the interagency project was starting up, I had neither used it to forecast future discovery rates nor verified its power; nor, to my knowledge, had anyone else. The discovery process model that we had developed grew out of the idea that each wildcat well has an area of influence that ranges from a maximum, when this area does not overlap the area of influence of any previously drilled well, to a minimum, when the well has been drilled at a site that is covered by the area(s) of influence of one or more previously drilled wells (Singer and Drew, 1976).

Working with this model at the level of the exploration play had presented us with a set of problems to solve. We believed that moving up to the scale of the Permian basin, where more than 20 plays occurred, would be quite another matter. To begin with, we knew that the wildcat drilling and discovery data were not in the form required for a discovery rate forecasting analysis that used the physical exhaustion model. The data presented a myriad of problems, ranging from poor to incomplete coverage in the field reserve estimates to a discontinuity in the definition of a wildcat well in the Petroleum Information Corporation well file that occurred between 1962 and 1963 (when it was changed from any well intended to discover a new field to wells intended to discover new pools and extensions of existing fields). The change in the definition was easy enough to uncover, but its origin never could be verified. The managers of the Petroleum Information Corporation would never confirm my conclusion that the discontinuity occurred at the point where they started realtime recording of the well data; all they would say was that they did not disagree with it. In addition, much of the data was just plain wrong.

Because we had spent a considerable amount of time constructing and working with our discovery process model, it was only natural that we would want to use it as a basis for the exploration function in the interagency study. This discovery process model was specified for individual field size classes, and it contained a parameter that precisely characterized the efficiency of exploratory drilling:

$$f = 1 - (1 - A/B)^c \qquad (6.1)$$

where

f = fraction of the fields that have been found within an individual size class
A = area that has been exhausted
B = effective basin area
c = discovery efficiency for the size class

The model was constructed in this form to connect the progressive physical exhaustion of the search area with the idea that larger oil and gas fields tend to be discovered earlier than smaller fields. The discovery efficiency parameter characterizes the physical exhaustion of the search area with reference to the outcome of a random search process (that is, $c = 1$). For a random exploration process, the number of fields discovered in the given size class per unit area exhausted remains constant throughout the total exploration history of a region; for example, if the ultimate effective basin area were 10,000 square miles and 100 fields existed in the size class at the start of the search process, we would expect 10 fields to be discovered as each 1000 square miles of the region was exhausted (Table 6.1, column 2; Fig. 6.1, lower graph).

If the exploration process for a given field size class is more efficient than random, a larger number of these fields would be expected to be found during the earlier stages of exploration (Table 6.1, column 3; Fig. 6.1, upper graph); for instance, if the discovery efficiency were 2.5 in our previous example, the exhaustion of the first 1000 square miles would producc 23.2 discoveries versus 10 discoveries for the random search case.

It is theoretically possible for a discovery efficiency to be less than 1 during the discovery history of a search area, but this case is of no relevance in the real world because it implies a steadily increasing discovery rate through the end of the exploration process. This leads to the ridiculous conclusion that, in the long run, drilling would terminate when the discovery rate reaches its peak! In the short run, however, a sudden geologic insight or chance fluctuation could cause the discovery rate to increase for a period of time. Such an increase is schematically diagrammed in Fig. 6.2.

When we tested our discovery process model by using discovery rate data from the Denver basin, examples of temporally increasing rates of discovery were encountered. In Fig. 6.3, the graph of the rate of discovery of fields in the 0.512-million- to 1.024-million-barrel size class displays an increasing rate of discovery between 1500 and 3000 square miles in the physical exhaustion of

Table 6.1. Expected Number of Discoveries per 1000 Square Miles Searched for Efficiencies of 1 and 2.5, Assuming 100 Undiscovered Fields Exist

Area searched (square miles)	Expected number of discoveries per 1000 square miles searched	
	Efficiency = 1 (random search)	Efficiency = 2.5
1000	10	23.2
2000	10	19.6
3000	10	16.2
4000	10	13.1
5000	10	10.2
6000	10	7.6
7000	10	5.2
8000	10	3.1
9000	10	1.5
10,000	10	0.3

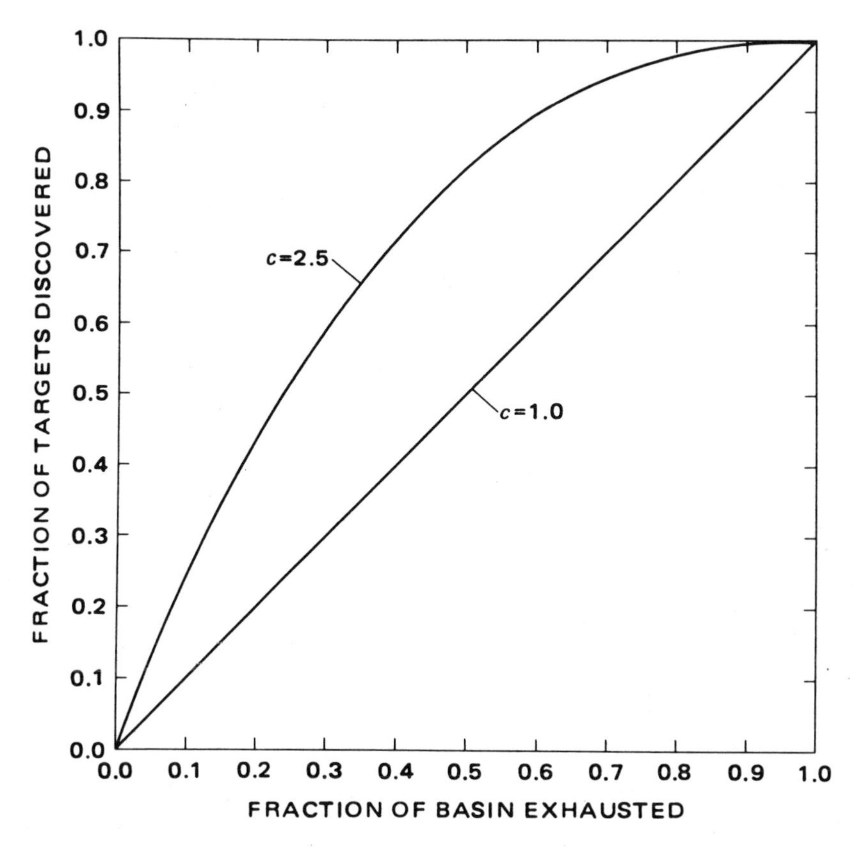

Figure 6.1. The fraction of the targets in a given size class discovered versus the fraction of the basin searched, given two different efficiencies of exploration C (Drew et al., 1980).

the D-J Sandstone stratigraphic unit in the basin. No obvious explanation could be found for this temporary increase in the rate of discovery.

In practice, forecasts of future discovery rates are only made for field size classes that normally have discovery efficiencies of 1 or greater; for example, the graph of the cumulative discovery rate for the 2.048-million- to 4.096-million-barrel size class in the Denver basin exhibits a markedly diminishing rate of return to the level of physical exhaustion of the search area (Fig. 6.4). The discovery efficiency is estimated to be 2.2 for the fields in this size class. The assessment of the future rates of discovery for all the size classes that have enough data to estimate the efficiencies of discovery were made and are displayed in Fig. 6.5. These results confirmed our earlier speculation that discovery efficiencies are higher for the larger field size classes; that is, larger fields are systematically discovered earlier than smaller fields in the exploration history of this basin.

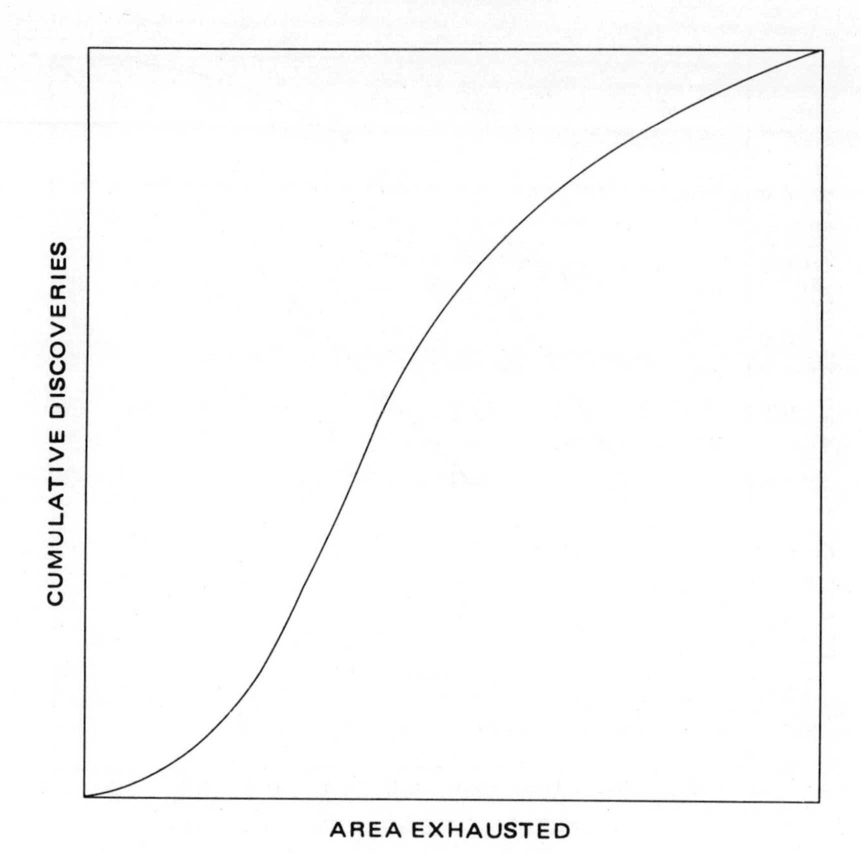

Figure 6.2. A temporarily increasing discovery rate (Drew et al., 1980).

The testing of our new discovery process model on the D-J Sandstone play in the Denver basin also included the estimation of the effective basin size before the ultimate number of fields expected to occur within each size class was estimated. The theoretical foundations for this estimation procedure were presented in Root and Schuenemeyer (1980).

A forecast of the outcome of a future increment of wildcat drilling in the Denver basin (actually a validated back-forecast) that we produced during the testing of the model is displayed in Fig. 6.6. To obtain a size-class-by-size-class comparison of the actual versus predicted discoveries of crude oil in the Denver basin for the period 1955–1969, we used data from 1954 and earlier to estimate the parameters.

When we pointed proudly to these results as validation of our discovery process model, the prevailing opinion was that it was an interesting but insignificant contribution to the large problem that the interagency task force was facing. There was also a feeling among some of the members that we were absolute crackpots to think that the data presented in Fig. 6.6 was a useful contribution.

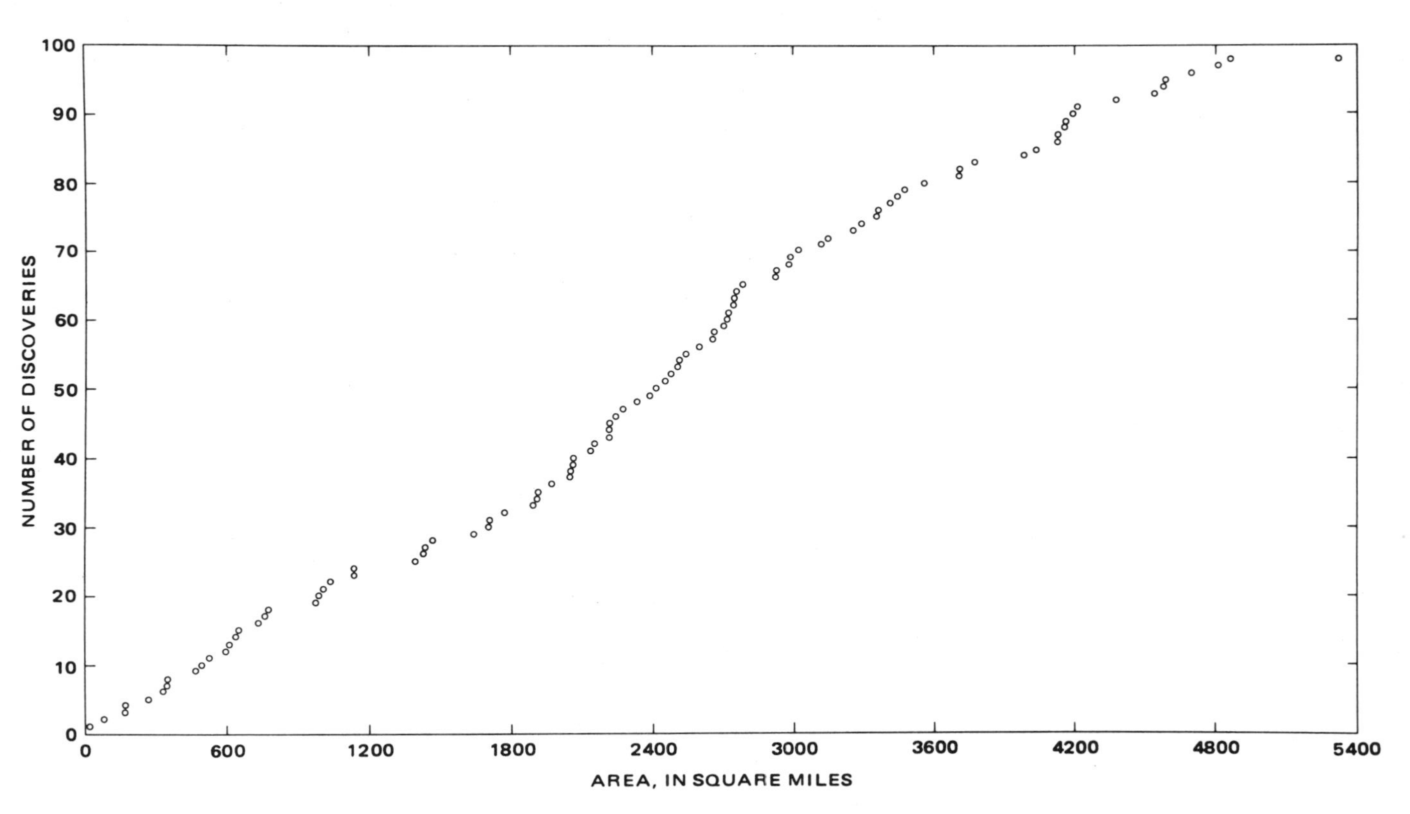

Figure 6.3. Cumulative discoveries versus area exhausted for class size 10 (0.512 million to 1.024 million barrels) (Drew et al., 1980).

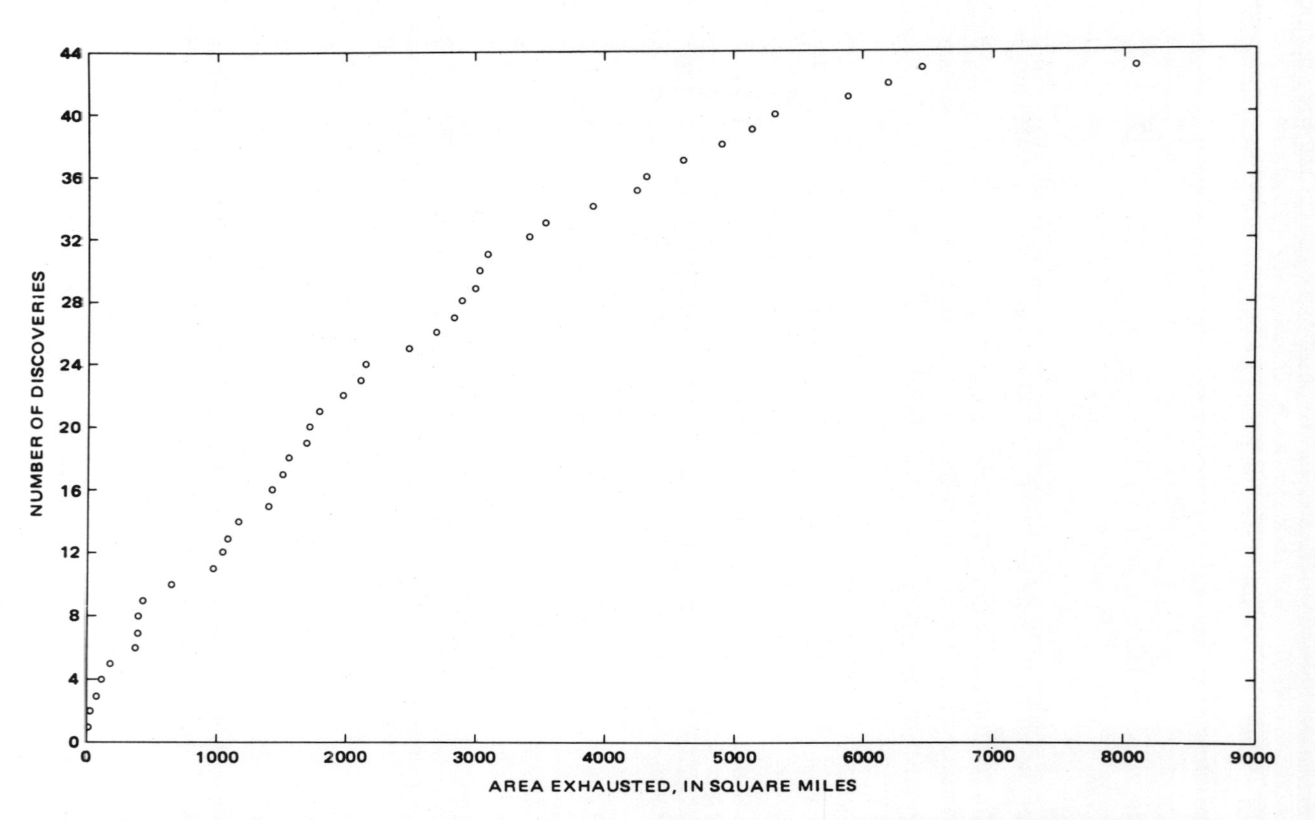

Figure 6.4. Cumulative discoveries versus area exhausted for class size 12 (2.048 million to 4.096 million barrels) (from Drew et al., 1980).

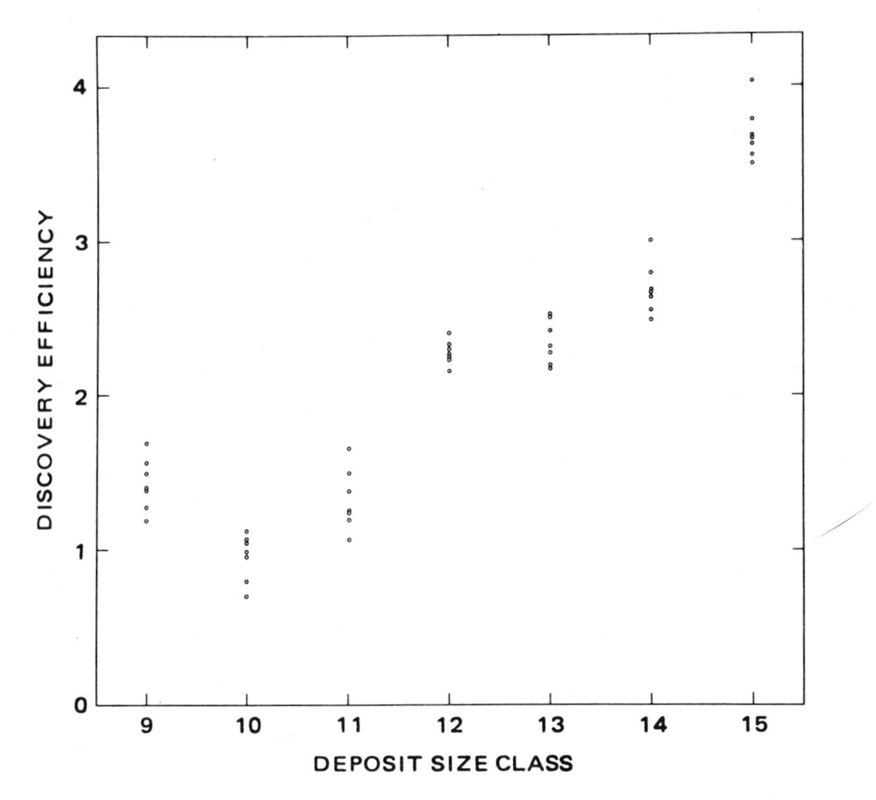

Figure 6.5. Computed discovery efficiencies for classes 9–15, 1963–1969 (Drew et al., 1980).

However, the type of forecast displayed in Fig. 6.6 partitioned onto a scale of wildcat wells drilled in the future was exactly what we needed so that engineers and economists could compute the marginal cost of converting the undiscovered 1976 oil and gas resources of the basin into future reserves by means of wildcat drilling. We knew that we faced the obstacle of identifying the individual fairways of the more than 20 exploration plays that had unfolded in the basin and then of assigning the past discoveries and wildcat wells to each of these plays. Our discovery process model also required a good deal of computer time to calculate the area exhausted by each new wildcat well drilled. The task seemed unachievable within the 6–9 months that had been allotted us to produce results.

If we had had more time, we could have proceeded in an orderly fashion toward our objective because the wildcat drilling data we needed for disaggregation of the exploration history of the basin into its individual exploration plays existed. The Petroleum Information Corporation well file for the Permian basin was touted as being one of their best regional subfiles. The Survey had

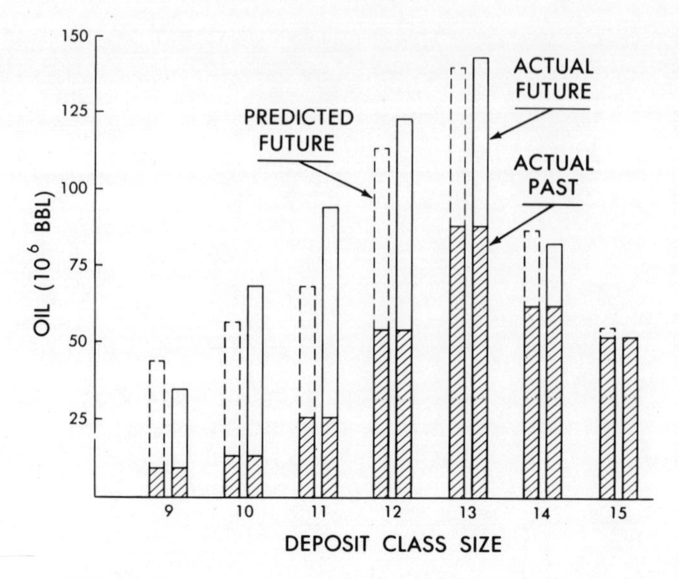

Figure 6.6. Comparison of actual versus predicted petroleum discovered within size classes 9–15 for discoveries made between January 1, 1955, and December 31, 1969, based on data to December 31, 1954.

purchased the file and had a big computer, so no procedural problems existed for accessing the drilling data.

In the Dallas field office of the EIA, the engineers, led by Tom Garland, said that they could produce ultimate productivity estimates for the more than 4000 oil and gas fields in the basin in six months. Although they could identify the fields by name and field code, railroad district, county, and depth, they were unable to identify a geologic label that could be used to separate these fields into exploration plays or the wildcat well that discovered the fields. However, the work they agreed to do was a large contribution to the interagency project.

Our desire to use our own discovery process model for the interagency study was soon set aside. The fuse on the interagency project had been set too short. I quickly came to the realization that it was with malice aforethought that Dick Meyer, the program coordinator, had set up the monthly meeting schedule. These meetings, which were like a metronome, were being used to marshall peer pressure against the six task leaders to keep up the pace. It was obvious that we were going to have to find a more efficient method or to make a compromise to forecast the future discovery rates in the Permian basin within the allotted time.

We discussed the use of a basin level decline extrapolation that would provide a forecast based on barrels of crude oil and thousands of cubic feet of natural gas per foot of future wildcat drilling. Such a forecast could be done quickly, but it would not provide the correct type of information for the economic forecasting model. This type of aggregate forecast lacked the critical element of producing a forecast in the form of incremental field size distributions as a function of various quantities of wildcat wells drilled in the future.

For no reason that I can remember, I took the 1958 paper by Arps and Roberts and scribbled out some Fortran code and compiled the program about a week before the third monthly interagency task force meeting. As I mentioned in the discussion of the contribution made by the Arps and Roberts model to the development of the exploration play concept in Chapter 4, I took the field discovery and wildcat drilling data from the Denver basin and, with no expectation of success, dumped it into the computer program. To my absolute amazement, the model made very good predictions. It was without design that I had chosen data from the same basin that Arps and Roberts had used for expositionary purposes. Here I was 20 years later coming along validating their model by using a back-forecast. Most importantly, I had done it in a few hours. The area of influence model that Jack Schuenemeyer, Dave Root, and I had developed and hoped to use in the interagency study had nice mathematical properties but required a lot of computer time to make a forecast. My immediate thought was that maybe the past discovery rate data for the Permian basin was regular enough that the Arps and Roberts discovery process model could be set up to make the required forecasts without the loss of too much accuracy to discontinuities and noise in the data. The next problem I had was how to tell the interagency task force about my discovery.

I filled my briefcase with computer output, made copies of the Arps and Roberts paper, and went enthusiastically to the next interagency meeting, which was being held in Denver, to announce that a solution might be at hand to the problem of forecasting the future rates of discovery in the Permian basin. I was ready to tell the members that, to make the great leap forward, I would need to make only one assumption, which was that the exploratory history of the Permian basin could be thought of as one big exploration play. When my turn came to tell of my progress that month, I would trot out the Arps and Roberts model and see if this assumption would pass the laugh test.

The meeting was well attended and went along in good order through the first five task group reports. Meyer juggled the schedule for each meeting, and this time my report was last. It was fortunate for the overall meeting that this was the order of business because when I started to talk about my new-found treasure, I ran into stiff opposition. One of the members of task group 1, which had the responsibility for the geologic assessment of the oil and gas resources, burst into a heated tirade. The task group 1 leader sat in quiet approval of the outburst. I decided to stand my ground. I recall one remark that these sorts of discovery process models give negative basin sizes. As I remember, I pointed out that the basin size is not computed endogenously in the model but, instead, is specified exogenously at the start of the analysis. It would not have mattered what I said. From then on, there was no chance to present my case formally. The meeting came apart at the seams. I could see that a reservoir of goodwill for discovery process modeling existed in the room from the odd glances. One member of the team from the DOE slid down almost out of sight in his chair, saying loud enough for all to hear that he could not believe what was happening. The wreckage was so complete that the program coordinator let things go until he could use the clock on the wall to call the meeting to a close. After it was over, I met him in the hallway and asked him what he wanted to do, given the heated and disparate reactions. He could have administered the opium of compromise, but he did not; he said simply, "Proceed."

The lesson I learned from coming out on the winning side of this confrontation was that it is a good policy to react slowly to somebody else's new idea. Trying to kill off a new idea simply because you do not like it often results in getting caught in your own snare. I remembered the advice of John Griffiths who said, "Watch your audience. If they clap politely, you have done absolutely nothing, but, if they hoot and howl, you are on the right track."

So, home we went with the Arps and Roberts model having been certified in a trial by fire and Dick Meyer's terse premission to proceed. The Arps and Roberts discovery process model is as follows:

$$F_A(w) = F_A(\infty)\left(1 - e^{-CAw/B}\right) \tag{6.2}$$

where

$F_A(w)$ = cumulative number of discoveries estimated to be made in size class A by the drilling of w wells
$F_A(\infty)$ = ultimate number of fields in size class A that occur in the basin
B = area of the basin
A = average areal extent of the fields in the given size class
w = cumulative number of wildcat wells, and
C = efficiency of exploration

To obtain a resource assessment in the form of a field size distribution directly from the drilling and discovery data, the model is solved backward for $F_A(\infty)$ for each field size class.

At the time we started to use this model, we did not realize that the resulting backward solution produced a biased estimate for the number of smaller oil and gas fields remaining to be discovered in the region under study.

A schematic diagram of the model is displayed in Fig. 6.7. Here we see that the larger fields are discovered more rapidly than the smaller fields; that is, $F_{A_4}(w)$ is a larger proportion of $F_{A_4}(\infty)$ than $F_{A_1}(w)$ is of $F_{A_1}(\infty)$. It should be noted that the exhaustion of the fields within each size class A goes as a declining exponential function.

Our first task was to determine if we could model the entire Permian basin as a single exploration play by using the Arps and Roberts model. We started by making a work file of wildcat wells out of the three reels of magnetic tape that the Petroleum Information Corporation had supplied to us from their

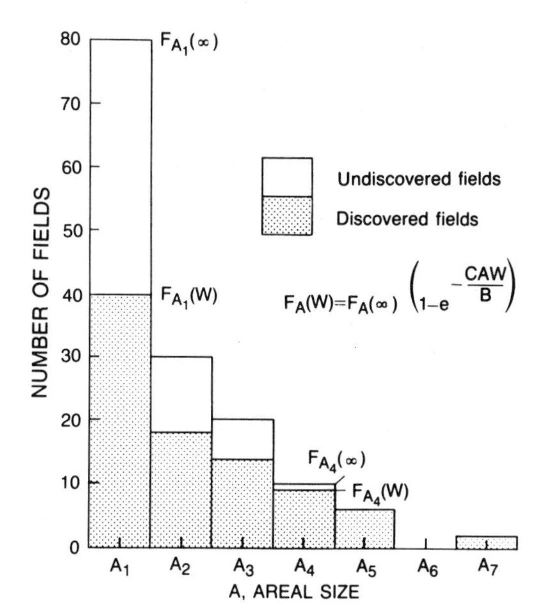

Figure 6.7. The Arps and Roberts discovery process model after *w* exploratory wells have been drilled.

master well file. There were 209,000 wells on these reels of tape. I wrote a computer program that classified these wells as wildcat, exploratory, or development and, given this classification, whether they were successful or not. This tabulation also was done for each of the counties in Texas and New Mexico within the Permian basin. We used the AAPG basin classification produced by a committee, led by Dick Meyer several years earlier, that used state and county codes as a basis for classification (Meyer et al., 1970, 1974).

The output from the computer program that made this tabulation was so large that it had to be pasted on a wall to be inspected. The major discontinuity in the wildcat well data that I mentioned earlier was immediately obvious by only a cursory observation of the computer printouts. Before 1963, nearly all the wells drilled were classified as having either predrilling intent of wildcat or production; that is, they carried predilling intent codes of 5 or 6, respectively. From 1963 onward, wells occurred in all the exploratory well classes, not just in the wildcat class. What had happened? Had there been a law that no one could drill a deeper or shallower pool test before 1963? Also, why were there so few outpost and extension wells? There had to be an explanation.

I called Leo Broin, an acquaintance who worked for Cities Service Oil Company, and told him what I had found. He knew a lot about the Petroleum Information Corporation well file, and, maybe, just maybe, he could tell me what I wanted to know. His first remark was, "It's just a scout ticket file!" He told me that the company that started the file sold out to another company, which, in turn, sold the file again.

I asked if the historical reconstruction process was stopped at 1962 and the realtime recording started in 1963. He said, "Maybe it could have been some year like that." He told me that it would not have been unreasonable during the historical reconstruction phase to have made such a decision to take all the wells and put them into two piles. One pile would contain the wells intended to discover new fields and pools and the reserves added through outpost and extension wells. The other pile would get all the direct offset wells and any other wells that clearly did not fall into the first pile. So, there I was with the best advice I could get—I had a definite maybe.

Soon after I talked to Leo Broin, a most unusual turn of events occurred. A telephone call came from the chief geologist's office asking me to come up and talk to one of the founders of the Petroleum Information Corporation. He was in town making calls and had stopped by to see how we liked his product. I was asked to take him to my office and to talk to him for an hour or so. I was greeted with the firm handshake and strong eye contact of the confident salesman. Back in my office he said, "How can I help you?" I took him to my new wallpaper, which was made of computer outputs of well classifications for the Permian basin from 1920 to 1976, and said, "Cast your peepers on this little glitch in your Permian basin file." He took some notes and asked a question or two. His watch told him he had to run. However, he promised that he would get right on my problem when he got back to Denver. He never called back nor did he have one of his staff call. More than 10 years have gone by, and I have never found anyone who would say with authority, "Yup, that's what happened back in 1963."

This discontinuity caused us quite a problem. When the ultimate productivity estimates for each field in the Permian basin were plotted by size class against the number of cumulative wildcat wells, the break in the 1962–1963 data really stood out. If you believe the graphs, it meant that a dramatic increase in the discovery rate occurred between these two years. This was nonsense! We pondered the question of how we were going to get around the problem. There was really nothing that could be done except to combine the wildcat wells with the four classes of exploratory wells after 1962 so that a consistent data series would exist as a basis for discovery process modeling. This adjustment would not be the last compromise made so that we could make a forecast of the future rates of discovery in the Permian basin. Before the project ended, Dave Root said that we had founded an exclusive club that he called "The Ruthless Smoothers and Approximators Club of America."

So, we would drive the discovery process model on a scale of total exploratory wells rather than on one of total wildcat wells drilled. We would have to assume that the relation between total exploratory and total wildcat wells would not change significantly in the future. Once this shift was made, we obtained very smooth graphs of the cumulative discovery rates within each field size class. We were elated. We then tried our first forecast.

Our first step was to estimate the parameters in the Arps and Roberts model by using the drilling and the discovery data from the entire basin through 1960. A back-forecast for 1961–1974 was made to check the model before a realtime forecast was made for the future starting in 1975. The total volume of oil and gas in barrels of oil equivalent was used as a criterion to estimate the parameters in the model.

When we examined the results of the back-forecast, we found them to be too pessimistic. The fields discovered between 1961 and 1974 inclusive were bigger on average than they should have been had the model been fitted correctly. The model had worked well in the simpler cases we had done. What could be going on in the Permian basin? In the simpler cases, such as the Denver basin, the exploration play was restricted to one stratigraphic unit. The Permian basin, however, was one in which more than 20 exploration plays had unfolded. The principal exploration targets in these plays ranged in depth from less than 2000 feet to more than 15,000 feet. A graph of the average depths of exploratory wells and discoveries showed a pattern of increasing depth through time. The question we asked ourselves was, "Could it be that the larger discoveries made in the basin during the 1961–1974 time interval were coming out of increasingly deeper and deeper depth intervals?" Consequently, the average size of the new discoveries was kept larger than it would have been had the explorationists not steadily sought out deeper and deeper targets during this time interval. Dave Root said he could imagine a matrix of partial derivatives set equal to zero describing the trade-off between the expected sizes of the fields remaining to be discovered in the different depth intervals and the average cost of exploration in each of these intervals. When he reached this conclusion, he remarked that there might be something useful in economic theory after all!

We then knew that we had to pull the exploration history of the basin apart by depth interval. From the distribution of discoveries by depth, we decided to use 4000-foot depth intervals. We got around the fact that many fields have multiple producing horizons by assigning to each field the depth interval at which its largest reservoir was found. The idea of the net exploratory well was concocted to solve the harder problem of how to assign an exploratory well to a depth interval. This idea is best explained by an example. Given that we were using, say, a 4000-foot depth interval and had a 10,000-foot well, we would declare that this well was one net well in the 0- to 4000-foot interval, one net well in the 4000- to 8000-foot interval, and one-half of a net well in the 8000- to 12,000-foot interval.

With this classification of exploratory drilling by depth interval, we returned to our forecasting efforts and recalibrated the Arps and Roberts model on the pre-1961 discovery and drilling data and then made a back-forecast for the period 1961–1974 to validate the model. We had accounted for the fact that the economics of exploration had caused the explorationists to make decisions to drill deeper during the 1961–1974 time interval to find larger oil and gas fields than they would have had they maintained a constant depth of exploratory drilling. We had a calibrated discovery process model that made reasonable predictions at the basin level. Dave Root and I were satisfied with the calibration and happy to have stumbled on the evidence in the drilling and discovery data of the trade-off that had to be made between the size of oil and gas fields and their depth.

We finished our forecast barely in time for Meyer's next monthly meeting. Because we had been talking to Emil Attanasi, the economist in our group and the petroleum engineers in the Dallas field office of the EIA about our efforts, our report was not going to be a total surprise. Apparently, we had not made enough of our choice of 4000-foot depth intervals because Tom Garland announced that it was nice that we had a usable forecast based on successive 4000-foot intervals but that that choice was not satisfactory to his group. He said that they had worked for more than 10 years to make cost estimates for drilling and equipping wells on 5000-foot depth intervals and that they were not going to redo their calculations because we could produce a better forecast using 4000-foot intervals. I momentarily objected that opening the depth interval would significantly reduce the precision and the accuracy of our forecast. The project coordinator looked at Garland, who said that to convert to another set of depth intervals would be a mountain of work they did not want to do. I quickly withdrew my objection.

Changing the depth interval was not the only compromise we made with the engineers. One of the most unusual quirks in the Permian basin report (Drew et al., 1979; U.S. Geological Survey, 1980) is the odd field size classes that we used (Table 6.2). These categories have been carried forward for consistency in all our reports in which we produced discovery rate forecasts. We have been quizzed repeatedly on why we used such a size classification. My answer is that it is what the engineers insisted that we use. The complete answer is buried somewhere in a decision made years ago in their cost-estimating procedures. If you press Tom Garland or any member of his team for an explanation, they will say that it had to do with the gas-to-oil ratios in crude oil that were used to make certain basic calculations. This classification of field sizes does not hurt anything—it is just an oddity that we now carry forward with each new forecast to maintain consistency with the last forecast.

After adapting to these changes, we were given the green light to go ahead with a final forecast (Drew et al., 1979; U.S. Geological Survey, 1980, pp. 20–28). The first step in the forecasting process was to estimate the ultimate number of fields in each field size class, $F_{A_i}(\infty)$, $i = 1, \ldots, 20$. To start this process, a value of C had to be estimated. The criterion for selecting a value of C for a particular depth interval was that the total oil and gas combined in the 1961–1974 forecasted discoveries be equal to the total oil and gas in the 1961–1974 actual discoveries.

The value of $F_{A_i}(\infty)$ was calculated for each size class of fields within each

Table 6.2. Field Size Classes

Size class	Size range (million BOE[a] recoverable oil and gas)
1	0.0 to 0.006
2	0.006 to 0.012
3	0.012 to 0.024
4	0.024 to 0.047
5	0.047 to 0.095
6	0.095 to 0.19
7	0.19 to 0.38
8	0.38 to 0.76
9	0.76 to 1.52
10	1.52 to 3.04
11	3.04 to 6.07
12	6.07 to 12.14
13	12.14 to 24.3
14	24.3 to 48.6
15	48.6 to 97.2
16	97.2 to 194.3
17	194.3 to 388.6
18	388.6 to 777.2
19	777.2 to 1554.4
20	1554.5 to 3109.0

[a]BOE = barrels of oil equivalent.

depth interval by solving the discovery process model backward. A sample calculation for field size class 10 in the 0- to 5000-foot interval is given below:

	Input data for size class 10; depth interval, 0–5000 feet	
A	Average areal extent of fields	2.2 square miles
B	Permian basin size	100,000 square miles
C	Efficiency of exploration	2.0
W	Cumulative exploratory wells through 1960	14,243
$F_A(w)$	Number of discoveries made in size class 10 in the 0- to 5000-foot interval through 1960	59

The solution for ultimate number of fields occurring is as follows:

$$F_{10}(\infty) = \frac{F_{10}(w)}{1 - e^{-CAw/B}} = \frac{59}{1 - e^{-2.0\times2.2\times14{,}243/100{,}000}} = 126.7 \text{ fields}$$

Given the number of discoveries in this size class through 1960 (59 fields) and the number of net wells drilled in the interval, the model estimates that 126.7 − 59 = 67.7 fields (expected) of this size (1.52 million to 3.04 million barrels of oil equivalent) remain to be discovered in the interval after 1960. This forecast can be checked against the actual number of discoveries made in this field size class from 1960 to any particular future year. The forecast shown below projects discoveries in size class 10 in the 0- to 5000-foot interval through 1974. On the basis of the 25,055 exploratory wells drilled in the basin through 1974, we have

$$F_{10}(25055) = 126.7\left(1 - e^{-2.0\times 2.2\times 25{,}055/100{,}000}\right)$$

= 84.5 fields to be discovered by December 31, 1974.

Because 59 discoveries had been made in this size class by 1960, the model is forecasting an additional 84.5 − 59 = 25.5 (expected) fields to be discovered in size class 10 between 1961 and 1974 inclusive. In fact, 25 discoveries actually were made in this size class during this 15-year period. Given the coarseness of the fitting criterion, such a result is very acceptable. The results of analogous calculations for all the field size classes in this depth interval are presented in Table 6.3. The level of accuracy varies across the size classes and also across the depth intervals, but acceptable agreement was found between predicted and actual levels of discoveries in nearly all cases; for example, the major differences between the actual and the predicted number of discoveries in the 0- to 5000-foot intervals were found to occur in six of the eight smallest class sizes (Table 6.3). In each of these classes, the model underpredicted the actual level of discovery. In the short run, the consequence of these underpredictions is not too significant because of the small amount of oil and gas contained in these fields. In the long run, however, we knew that, as these smaller fields became increasingly more important, better techniques would have to be used to calibrate the model.

The actual forecast given to the economist and the engineers to be used in building the marginal cost curve for future discoveries was made by employing the discovery and the drilling data through 1974 to estimate the parameter in the Arps and Roberts discovery process model. The ultimate number of fields estimated to occur in each size class within each depth interval is used in the model to forecast the future rates of discovery in realtime starting at the beginning of 1975 (Tables 6.4–6.7).

This forecast was made for 20 successive increments of 1000 exploratory wells drilled at the surface of the basin. Within each of these drilling increments, a portion of the wells was applied, on a net well basis, against the expected size distribution of fields remaining to be discovered in each depth interval at the start of drilling. The assignments of the net wells to be drilled in each drilling increment were taken from extrapolations of the historical trends in the net well penetration in each depth interval versus the cumulative wells

Table 6.3. Comparison of the Numbers of Discoveries Through 1974 with the Number Forecast, Keyed on the Pre-1961 Discovery and Exploratory Drilling Data in the 0- to 5000-Foot Depth Interval

		Discoveries		
Size class	Area (square miles)	Actual	Estimated	Difference
1	0.13	211	200.1	10.9
2	0.14	84	66.0	18.0
3	0.16	84	84.7	−0.7
4	0.22	92	67.0	25.0
5	0.25	127	109.6	17.4
6	0.39	135	114.8	20.2
7	0.49	129	125.3	3.7
8	0.94	104	92.8	11.2
9	1.23	111	107.4	3.6
10	2.21	84	84.5	−0.5
11	4.24	75	82.9	−7.9
12	5.40	49	52.2	−3.2
13	8.30	35	33.3	1.7
14	18.19	19	18.2	0.8
15	40.42	14	14.0	0.0
16	49.19	16	16.0	0.0
17	67.20	9	9.0	0.0
18	81.75	6	6.0	0.0
19	129.88	2	2.0	0.0
20	40.25	1	1.0	0.0
Total		1387	1286.8	100.2

Table 6.4. Expected Ultimate Number of Fields Remaining To Be Discovered After 1974 in the 0- to 5000-Foot Depth Interval

Size class	Area (square miles)	Expected number of fields remaining	Number of fields found by 1974
1	0.13	4106.1	211
2	0.14	1355.3	84
3	0.16	1006.3	84
4	0.22	789.4	92
5	0.25	951.6	127
6	0.39	625.5	135
7	0.49	463.5	129
8	0.94	175.2	104
9	1.23	130.3	111
10	2.21	41.4	84
11	4.24	10.2	75
12	5.40	2.5	49
13	8.30	0.1	35
14+	46.18	0.0	67

Table 6.5. Expected Ultimate Number of Fields Remaining To Be Discovered After 1974 in the 5000- to 10,000-Foot Depth Interval

Size class	Area (square miles)	Expected number of fields remaining	Number of fields found by 1974
1	0.12	5462.9	215
2	0.12	2922.0	115
3	0.13	2856.8	122
4	0.19	2189.6	138
5	0.20	2468.0	164
6	0.29	1718.2	168
7	0.42	1382.6	200
8	0.47	751.6	160
9	0.76	570.5	158
10	1.28	245.4	125
11	2.21	94.6	98
12	3.76	26.6	76
13	5.40	5.9	50
14	10.86	1.5	47
15	20.30	0.2	27
16+	72.14	0.0	27

Table 6.6. Expected Ultimate Number of Fields Remaining To Be Discovered After 1974 in the 10,000- to 15,000-Foot Depth Interval

Size class	Area (square miles)	Expected number of fields remaining	Number of fields found by 1974
1	0.21	511.0	49
2	0.22	307.9	31
3	0.33	277.2	36
4	0.25	400.7	56
5	0.31	320.6	48
6	0.37	302.7	53
7	0.31	325.4	62
8	0.43	249.1	54
9	0.63	196.1	62
10	1.30	97.8	47
11	1.20	93.1	64
12	1.90	34.5	48
13	3.89	6.3	28
14	4.77	1.6	21
15	7.53	0.3	18
16+	12.67	0.0	9

Table 6.7. Expected Ultimate Number of Fields Remaining To Be Discovered After 1974 in the 15,000- to 20,000-Foot Depth Interval[a]

Size class	Area (square miles)	Expected number of fields remaining	Number of fields found by 1974
1	1.00	17.6	2
2	1.00	17.6	2
3	0.00	0.0	0
4	1.00	26.5	3
5	1.00	17.6	2
6	0.78	35.3	4
7	1.00	17.6	2
8	0.25	8.8	1
9	1.00	8.8	1
10	2.25	22.9	4
11	3.67	10.7	3
12	4.00	1.9	1
13	4.50	2.7	2
14	9.00	2.6	3
15	10.50	5.5	9
16	9.00	0.3	1
17	21.00	0.2	1
18	14.00	0.1	1

[a]The USGS resource appraisal group estimated that no oil and a mean value of 2.29 trillion cubic feet of gas in place exists below 20,000 feet.

drilled at the surface through 1974. A mathematical function was fitted to each of these net-well curves and then extrapolated for an additional 20,000 exploratory wells to be drilled at the surface in the future.

The number of discoveries expected to be made within each field size class for any future increment of exploratory drilling was determined by aggregating the expected number of discoveries to be made within each depth interval. The first increment of 1000 exploratory wells drilled at the surface results in 900 net wells in the 0- to 5000-foot interval, 500 net wells in the 5000- to 10,000-foot interval, 130 net wells in the 10,000- to 15,000-foot interval, and 35 net wells in the 15,000- to 20,000-foot interval.

Dave Root and I forecast that within the 1000-well increment, 150.9 discoveries would be made (Table 6.8). In the sense of having a nontrivial expectation, the largest discovery predicted in this first drilling increment is in size class 17 (194.3 million to 388.6 million barrels of oil equivalent). This discovery (in expectation, 0.1 field) is predicted for the 15,000- to 20,000-foot interval. With the exception of size class 1, the largest number of discoveries expected within a single size class is predicted to occur in size class 7 (0.19 million to 0.38 million barrels of oil equivalent) where 16.7 fields are expected to be discovered within the four depth intervals.

Comparison of the forecast of the expected number of discoveries across the

Table 6.8. Number of Discoveries Expected To Be Made in Each Size Class with the First Increment of 1000 Exploratory Holes Drilled in the Permian Basin after 1974

Size class	0–5000 foot depth	5000–10,000 foot depth	10,000–15,000 foot depth	15,000–20,000 foot depth	Total
1	7.5	10.1	2.8	0.3	20.7
2	3.0	5.4	1.7	0.3	10.4
3	2.9	5.7	2.0	0.0	10.6
4	3.2	6.4	3.1	0.4	13.1
5	4.3	7.6	2.6	0.3	14.8
6	4.4	7.7	2.9	0.5	15.5
7	4.1	8.9	3.4	0.3	16.7
8	3.0	6.9	2.9	0.1	12.9
9	2.9	6.6	3.2	0.1	12.8
10	1.6	4.8	2.3	0.5	9.2
11	0.8	3.2	2.8	0.4	7.2
12	0.3	1.7	1.7	0.1	3.8
13	0.0	0.6	0.6	0.2	1.4
14	0.0	0.2	0.2	0.3	0.7
15	0.0	0.1	0.1	0.7	0.9
16	0.0	0.0	0.0	0.0	0.1
17+	0.0	0.0	0.0	0.1	0.1
Total	38.0	75.9	32.3	4.7	150.9

depth intervals revealed a trend in the forecasted size of the future discoveries. During the first drilling increment of 1000 exploratory wells drilled at the surface starting in 1975, no discoveries were forecast to be larger than field size class 12 (6.07 million to 12.14 million barrels of oil equivalent) in the shallowest, most explored depth interval; discoveries were expected to be made up to size class 15 in the second and third depth intervals; and discoveries were expected to be made in size classes 16 and 17 in the deepest depth interval. The total volume of petroleum contained in the 150.9 fields expected to be discovered in this first drilling increment was estimated to be 254.1 million barrels of oil equivalent (Table 6.9, column 6, row 1,). The total volume of petroleum expected to be discovered in each of the 20 drilling increments along with the individual contributions made by the expected discoveries in each depth interval are shown in this table.

In terms of the assessment of the undiscovered petroleum resources of the Permian basin, the interagency study (U.S. Geological Survey, 1980) concluded that a large number of oil and gas fields remained to be discovered in the basin—approximately 34,000 fields at depths shallower than 20,000 feet—but most of these fields individually contain very small volumes of oil and gas (Table 6.10). The model predicted that only 708.9 of these fields (2.1 percent) individually contained more than 1.52 million barrels of oil equivalent of reserves.

Furthermore, the model predicted that nearly all the 6.6 expected fields remaining to be discovered in the basin that contained more than 48.6 million

Table 6.9. Total Volume of Petroleum Predicted To Be Discovered (Millions of Barrels of Equivalent) by Depth Interval in 20 Successive Increments of 1000 Wells Each of Future Exploratory Drilling in the Permian Basin

Drilling increment	Depth interval (in thousands of feet)				
	0–5	5–10	10–15	15–20	Total
1	16.5	74.8	61.0	101.8	254.1
2	15.7	71.2	58.2	91.4	236.5
3	15.0	67.9	55.5	81.2	219.6
4	14.4	64.9	53.1	71.6	204.0
5	13.8	62.2	50.8	62.7	189.5
6	13.2	59.7	48.7	54.5	176.1
7	12.7	57.4	46.7	47.2	164.0
8	12.2	55.2	44.9	40.6	152.9
9	11.7	53.2	43.1	34.9	142.9
10	11.3	51.4	41.4	29.9	134.0
11	10.9	49.6	40.0	25.5	126.0
12	10.5	48.0	38.4	21.8	118.7
13	10.1	46.4	37.0	18.6	112.1
14	9.8	45.0	35.7	15.8	106.3
15	9.5	43.7	34.4	13.5	101.1
16	9.2	42.4	33.2	11.6	96.4
17	8.9	41.2	32.1	9.9	92.1
18	8.6	40.1	30.9	8.5	88.1
19	8.4	39.0	29.9	7.4	84.7
20	8.1	38.0	28.8	6.4	81.3
Total	230.5	1051.3	843.8	754.8	2880.4

barrels of oil equivalent each were expected to occur in the 15,000- to 20,000-foot depth interval. In contrast, the 20.7 fields remaining to be discovered that individually contained between 12.14 million and 48.6 million barrels of oil equivalent were expected to occur in approximately equal numbers in the 5000- to 10,000-foot, 10,000- to 15,000-foot, and 15,000- to 20,000-foot depth intervals.

A forecast for each of the drilling increments, such as that displayed in Table 6.8, was the final product of our work. The data in these tables were used by

Table 6.10. Tabulation of the Number of Oil and Gas Fields Larger than 1.52 Million Predicted To Be Remaining at the Start of 1975 in the Permian Basin

Field size (millions of barrels of oil equivalent)	Number	Percentage of total remaining fields
48.6 and larger	6.6	0.02
12.14–48.6	20.7	0.06
1.52–12.14	681.6	2.00

Source: U.S. Geological Survey, 1980.

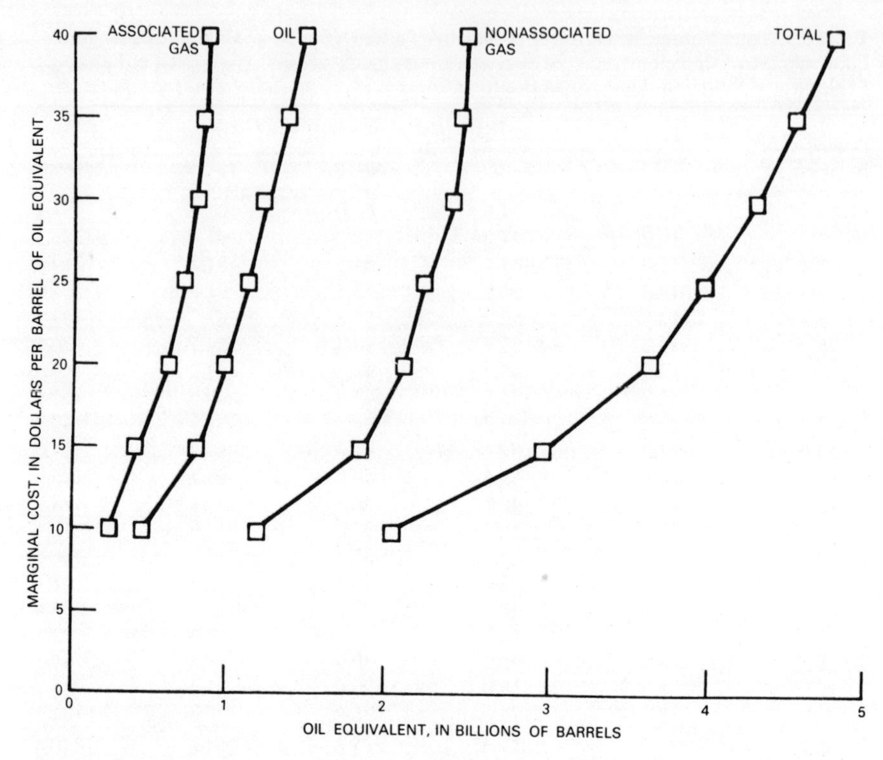

Figure 6.8. Marginal cost of recoverable oil and gas resources from undiscovered fields in the Permian basin—5-percent discounted cash flow rate of return.

the economist and the petroleum engineers to construct the economic model that turned this forecast into suites of marginal cost curves (Figs. 6.8–6.10). It is curves of this type that can be used to answer the question that we were asked to solve as a consequence of the follow-up study to Circular 725. It had been asserted by Hermann Enzer, the economist who composed those infamous words in the introduction, that a price rise for oil and gas, such as that experienced in the early 1970s, would cause the size of the economic undiscovered resources to increase and that this amount could be calculated.

Enzer may have accepted as a given that price rises in raw materials, like crude oil, cause such increases. His real point was that we should get on with the business of calculating their magnitudes. I will not go into the complexities of how a rising price causes different parts of the underlying field size distribution to become targets for exploratory drilling because this is onc of the topics of Chapters 8 and 9. Instead, I will say that marginal cost curves, such as those shown in Figs. 6.8–6.10, express the consequences of price movements on the quantity of a resource that is economically available, given an estimate of the size distribution of undiscovered fields and certain cost conditions.

In the summary of the results of the interagency study (U.S. Geological Survey, 1981, pp. 40–41), it was concluded, "If the price is assumed to be $40 per BOE [barrels of oil equivalent], the economically recoverable oil equivalent attains a maximum of 4.7 billion BOE at a 5 percent rate of return, 4.3 billion BOE at a 15 percent rate of return, and 3.9 billion BOE at a 25 percent rate of return. These quantities can be compared with the 38.2 billion BOE already discovered in the Permian basin by the end of 1974."

Between 30,000 (at a 25-percent rate of return) and 48,000 (at a 5-percent rate of return) exploratory wells were estimated to be economically viable at the $40 price level. This can be compared with the slightly more than 30,000 exploratory wells that were drilled in the Permian basin through December 31, 1974, where approximately 10 times more barrels of oil equivalent were discovered than that projected for the next 30,000–48,000 wells. At $15 per barrel and a 15-percent rate of return, only 12,000 exploratory wells would be drilled, and 2.3 billion barrels of oil equivalent reserves would be added.

The model can be used to answer two questions. First, what price and rate of return are required to stimulate a certain level of exploration to find a given

Figure 6.9. Marginal cost of recoverable oil and gas resources from undiscovered deposits in the Permian basin—15-percent discounted cash flow rate of return.

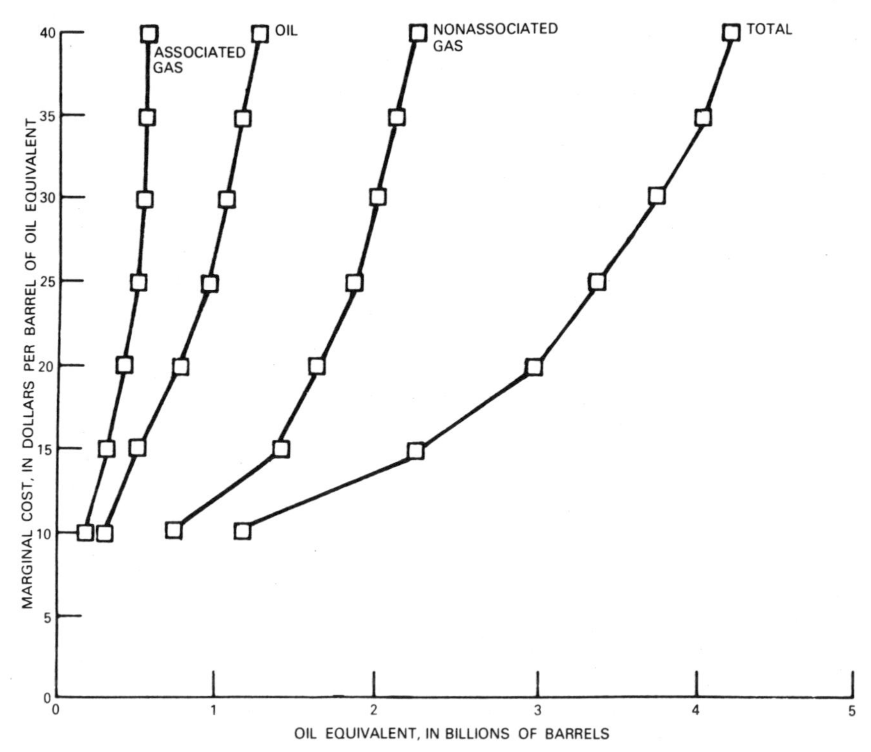

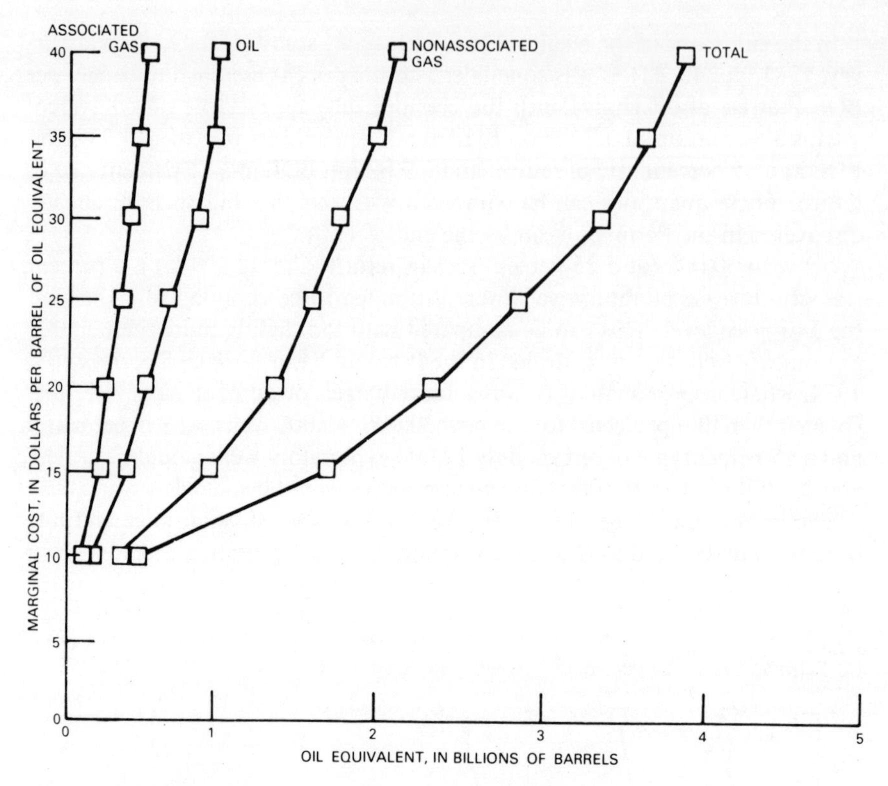

Figure 6.10. Marginal cost of recoverable oil and gas resources from undiscovered deposits in the Permian basin—25-percent discounted cash flow rate of return.

amount of oil and gas in barrels of oil equivalent? Second, what amount of oil and gas can be anticipated from the Permian basin at the present or some future price and rate of return?

The output from the model can be examined in terms of associated-dissolved gas, nonassociated gas, and oil resulting from various price levels and rates of return. The graphs shown in Figs. 6.8–6.10 reveal that the declining size of the targets discovered is a result of additional exploratory effort. As the marginal cost curves turn sharply upward, oil- and gas-reserve additions to be gained at costs of $10–$25 per barrel are considerably greater than those of $25–$40 per barrel.

Another result shown in the analysis is the dominance of the nonassociated gas in the remaining resources of the Permian basin. This dominance is true under all price and rate-of-return assumptions. When nonassociated gas is added to the associated-dissolved gas, crude oil represents only about one-quarter of the future potential.

With the draft reports from each task group in hand, Dick Meyer declared the Interagency Oil and Gas Supply Project a success at the start of what we suddenly realized was to be our last monthly meeting. A suggestion was made by a member to carry on with the second phase of the project as originally outlined by moving to the Gulf of Mexico, which was to serve as a case study for "a partially developed basin." According to the original outline, the Permian basin had been the mature case, and the Baltimore canyon had been intended to be used as the frontier case study. It was decided that the second phase would be carried out in a research project that had no rigid timetable and that the individual task leaders would work together as needed to get the job done. We all looked at each other, realizing that this really was the end of the interagency project. It was early spring 1978. We decided among ourselves that we would carry on with the Gulf of Mexico study, and the principal players would be the discovery process modelers along with Tom Garland and his petroleum engineers. We soon would be joined in our effort by John Haynes of Global Marine Inc. who made a very important contribution to our Gulf of Mexico study.

7

Discovery Rate Forecasting, Part 2: The Gulf of Mexico Offshore

With the completion of the Permian basin report in 1978, the work of the Interagency Oil and Gas Supply Task Force was finished. The working relations between the discovery process modelers and the engineers in the Dallas field office of the EIA were maintained on an informal basis through the completion of the second phase of work, which was forecasting the supply of oil and gas in the Gulf of Mexico, as set forth in the original task force agreement. Forecasting the future supply of oil and gas from the Gulf of Mexico offshore region turned out to be more involved than a simple variation on the theme that had been established by forecasting the future rates of discovery in the Permian basin. We took the time to improve our forecasting techniques and the collection and analysis of the wildcat drilling data.

As we started our work on the Gulf of Mexico region (Fig. 7.1), we were given some valuable advice—we should choose some other area where we would not have to deal with the confounding problems caused by the vagaries of the federal lease sale process, the disputes between state and federal ownership, and the trend in exploring in increasingly deeper water depths through time; each was believed to have introduced large inhomogeneities into the discovery data. The shift of exploration from oil toward natural gas in the early 1970s also was judged to be messy to untangle. Faced with these considerations, we could have either chosen another partially explored onshore area to study or agreed with the project coordinator that the interagency project, either

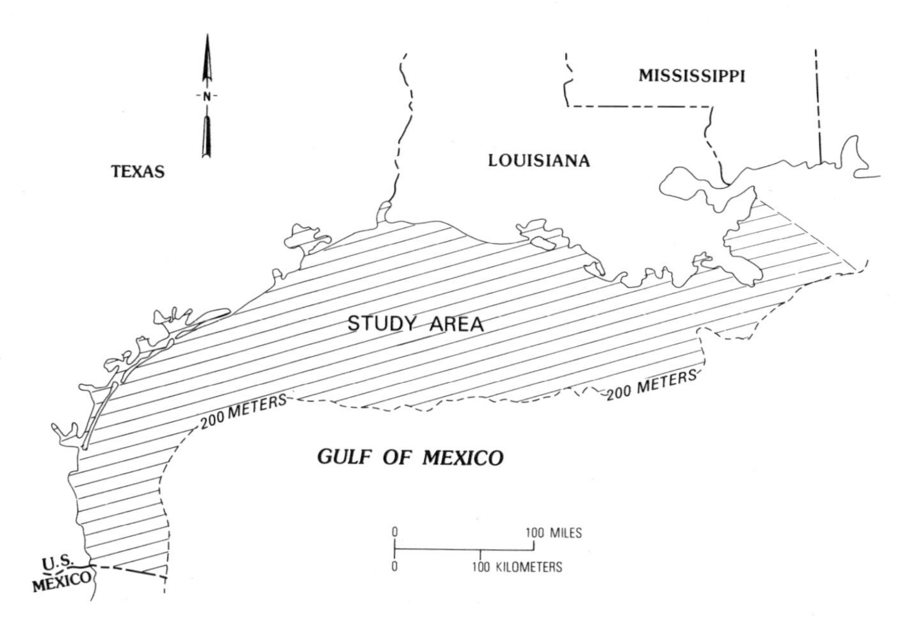

Figure 7.1. Location of our study area in the Gulf of Mexico.

formally or informally constituted, had gone on long enough and was declared to be a complete success with the completion of the Permian basin study.

We did neither. Instead, we stuck to the original plan. In hindsight, this turned out to be an important decision. The Gulf of Mexico was not only a large geographic area that had a large oil- and gas-resource base, but it was also a relatively high-cost area in which to explore and develop oil and gas fields. Also, high-quality drilling data were available in the public record and similarly high-quality field reserve data (proprietary) were available to us as members of the Geologic Division of the USGS from our Conservation Division, which is now the Minerals Management Service. Studying such a region was the key ingredient we needed to start the chain reaction of events that led to our understanding of how the oil and gas field size distributions we observed are built up by the discovery process over time. The scheme we developed to estimate the form of the underlying field size distribution in a partially explored area was a consequence of the manner by which sequential discoveries were added to the observed field size distribution in regions that had different cost regimens.

As we started this project, the attitude of "management" was, "Let's see if you make some progress on the data collection part of the project and then take it from there." The implication was that, if we could make a forecast of future discovery rates, the manpower and resources would be forthcoming to do the engineering and economic analysis necessary to produce the marginal cost curve for future discoveries in this offshore region. The first task to be completed was to build a wildcat well and field discovery data file for the Gulf of Mexico offshore. A USGS colleague, Walt Bawiec, and I started the data collection process in the autumn of 1978.

From the start of this study, we believed that we could improve our forecasts of future discovery rates of oil and gas by collecting and using better quality drilling data. We also believed that the discovery process model must be calibrated by a past series of wildcat wells, thereby allowing the forecast of future discovery rates to be driven by a stream of wildcat wells. We wanted to prevent having to make compromises, such as using exploratory wells (as in the Permian basin study), any partial aggregation thereof, or any other confounded drilling data series. Our plan was to collect the drilling data ourselves by using the most basic sources of data possible. These would include the International Oil Scouts Association *Yearbooks,* the annual "North American Development Issues" of the *Bulletin of the American Association of Petroleum Geologists,* the petroleum information well file of the Petroleum Information Corporation, and any other documents that we could find that displayed or listed the location and the characteristics of the wildcat wells drilled in the Gulf of Mexico. During this process, we found the Transco Pipeline Company maps to be most useful for cross-checking lease numbers and for determining well locations. We also used many of the studies of individual oil and gas fields in the Gulf of Mexico; these are most often published by local geologic societies (such as, Lafayette Geological Society, 1973). These studies almost always contain maps that not only have the oil and water contact plotted, but also the wildcat, exploratory, and development wells plotted and labeled. A six-volume series entitled *Opportunities for Increasing Natural Gas Production in the Near Term,* which was published by the National Research Council of the National Academy of Sciences, is an excellent example of these types of studies for the Gulf of Mexico offshore area. The third volume in this series was useful to us because, in addition to displaying data on wells, it was the only document available in which the unitized reserve figures of Eugene Island Block 266 and South Marsh Block 108 fields were separated into individual units so that these fields could be assigned correctly to the discovery well sequence. A large degree of trust can be placed in these types of data because they almost always are prepared by the geologists who were on the scene at the time the wells were drilled. Very often, such field studies are published by the geologist who made the basic geologic interpretations that resulted in the drilling of these wells. Knowing this, we relied on these as prime sources to obtain and validate well names, locations, total depths, and so forth.

We started collecting data by encoding the exploratory wells drilled in the Gulf of Mexico offshore region from the lists presented in Part 1 of the International Oil Scouts Association *Yearbooks* from 1945 to 1977. This was the most comprehensive and reliable source of drilling data. We collected our own data first-hand from the most basic sources to avoid as best we could any criticism from the energy analysis community, which was led by statisticians and operations research analysts in the newly formed EIA, that old-line government agencies and the oil and gas professional societies and trade associations were not objective when it came to data collection and analysis.

In the late 1970s, the members of this community believed that, over the years, the oil and gas industry trade associations and professional societies had become agents of the industry and were, therefore, antagonistic to the interests of the general public. I cannot keep from pointing out that most of these analysts migrated to the energy analysis field because it was the most action-packed place to be at the time. I believe that these men and women were driven by the idea that gaining power over the field in general meant gaining control of the data. If this is true, then it would be expected that they would try to discredit those who controlled the data.

Because of our involvement with oil and gas well and field reserve data, my colleagues and I attended many meetings at the DOE and the EIA and on Capitol Hill. The major topic of discussion usually was the integrity of the existing data files. We were often confronted by one or more of these "johnny-come-lately" energy analysts who would, for example, show or talk about lists of field names and field codes from different sources that did not match. The conclusion had been made that the data collectors were not capable of telling the truth or had conspired not to do so. Many times I wanted to tell these analysts that if they wanted to be sure of a good data set, they would have to start with a blank piece of paper, fill it up, and cross check as they went along; when the data set was complete, then they could stop. Sometimes, however, I could not remain quiet—I remember telling a high-ranking official in the EIA to stop asking me to 2 P.M. Friday meetings if all he planned to do was fuss about something that he was unwilling to try to understand.

Because of their attitude toward earth scientists, dealing straight ahead with these chaps was always an odd business. "They are from the Interior Department," the analysts would say to each other as they tried to excuse and/or understand our behavior. When it came down to it, they concluded that, if you were from Interior, you were different. Once we were so labeled, it was obvious that it served no purpose to say to these folks that oil and gas field data collection and compilation is a thankless job that has been done, for the most part, by altruistic people wherever they worked. We had to wait two or three years until the energy analysis scene cooled down as a consequence of the decline in federal funding for energy studies. When it did, many of these analysts moved on to the next hot spot where federal monies were flowing more freely. Those that stayed on matured into their jobs as they started to grapple sincerely with the data acquisition and analysis problems in the field of oil and gas supply analysis.

As a result of being involved in the energy analysis field during the middle and late 1970s, I learned two things from the prevailing wisdom—that the procedures used to prepare data series and subsequent analyses must have audit trails and that all data must be validated by modern methods. As best as I could determine, having an audit trail meant that anyone (say, any Ph.D. trained as a scientist or an economist) should be able to follow, to his or her satisfaction, any aspect of the analysis by expending only a minimum of effort and without any specific foreknowledge of the oil and gas industry. It seems to me that the important point about the audit trail is that it was necessary to have one in place so that it could be challenged. When an analyst asked for an exposition on the audit trail of the construction of a data file, it was a preemptive legal challenge. The best defense was to demonstrate a nonreliance on input from people who had specific knowledge about the oil and gas industry. These energy analysts knew that they had to get around the historical and the cultural aspects of the data-collection procedures, the jargon, and the obstacles encountered that revolved around issues in the earth sciences and the engineering aspects of the oil and gas business. They had neither the time nor the inclination to learn the organization of the oil and gas industry, so they engaged in quasi-legal procedural manueverings to put themselves in positions of power in the energy policy analysis field.

Given this climate, I was determined that we would leave an audit trail. In addition, we would put it right out in front, on page 1 of the report! The format of the exploratory well and discovery data file and the data from a sample well are given in Table 7.1. Our file contained information on a total of 7077 explor-

Table 7.1. Layout of the Data File on Exploratory Wells Drilled During 1945–1978 Offshore Texas and Louisiana in the Gulf of Mexico

Columns	Data	Data for example well
1–11	Subarea	High IS[a]
12–19	Well number	1 OCS
20–27	Lease number	1,831
25–31	Block number	206
32	Block specification	L[b]
33–38	Total vertical depth (feet)	8,587
39–40	Lahee class	50[c]
41–42	Completion date (month)	9
43–44	Completion date (a year between 1945 and 1978)	68
45	Geologic region	1[d]
46–47	Federal lease sale number	22[e]
48–57	Producible oil discovered (millions of barrels of oil)	0.0
58–66	Producible gas discovered (billions of cubic feet)	0.0
67–72	Area of field (square kilometers)	0.0
73–80	Operator	Humble

[a]Designation for "High Island area."

[b]Designation for "large block."

[c]Designation for "dry and abandoned wildcat well."

[d]Designation for "Miocene trend."

[e]Designation for "Federal lease sale number 22" (which was held on April 21, 1968).

atory wells, of which 4639 were wildcat wells and 2438 were other classes of exploratory wells. We went to some lengths to ensure that the purpose of a well labeled as a wildcat well had been to discover new oil and gas fields and that the purpose of other exploratory wells had been to extend already existing fields or to discover new pools within them. Emphasis also was placed on collecting data that were useful for the study of how geologic and institutional factors affected rates of discovery. We knew that the Gulf of Mexico offshore might present special problems in discovery rate analysis and forecasting that were related to the federal lease sale procedures and state versus federal jurisdictional disputes, a trend to explore in deeper and deeper water over time, and the switch from exploring principally for crude oil to exploring for natural gas. That is why the subregion, lease number, lease sale designation, operator, and geologic region data were encoded in addition to the completion date data. We knew beforehand that being able to examine the temporal rates of success in the Gulf of Mexico offshore as a function of institutional and geologic factors would be required before any analysis could be declared complete. Trying to model the diminishing rates of return to wildcat drilling as a basis on which to forecast the future rates of return would be a meaningless activity if the effects of institutional factors, such as the federal lease sale procedures, were not accounted for in the analysis.

Locating each exploratory well was conditional on the following: First, each exploratory well had to be located within the proper geologic trend (Miocene, Pliocene, or Pleistocene). Fortunately, a geologic trend map that had block and lease numbers had been prepared for Outer Continental Shelf (OCS) lease sale No. 38 by the then Conservation Division of the USGS, Metairie, Louisiana (Fig. 7.2). Second, a lease sale number had to be assigned to each well. Our source document for this, which was provided by the then Conservation Division, also was used to cross check and assign missing lease numbers. The lease numbers for the wells drilled in the state waters of the Gulf of Mexico were obtained, for the most part, from the International Oil Scouts Association *Yearbooks*.

The encoding sequence used to build the data file well by well was as follows:

1. The data were coded according to the format shown in Table 7.1 for each exploratory well listed in Part 1 of the International Oil Scouts Association *Yearbooks* for the period 1945–1978. These yearbooks were the most comprehensive source of drilling, discovery, and leasing information.

2. In June 1979, a retrieval to the same format was made from the Petroleum Information Corporation well history control file. This source was our second most important source of information. It was used to check wells listed and to add wells not listed in the first source. All the data for wells drilled in 1978 were obtained from this file.

3. The "North American Development Issues" of the *Bulletin of the American Association of Petroleum Geologists* were used to check and to add any

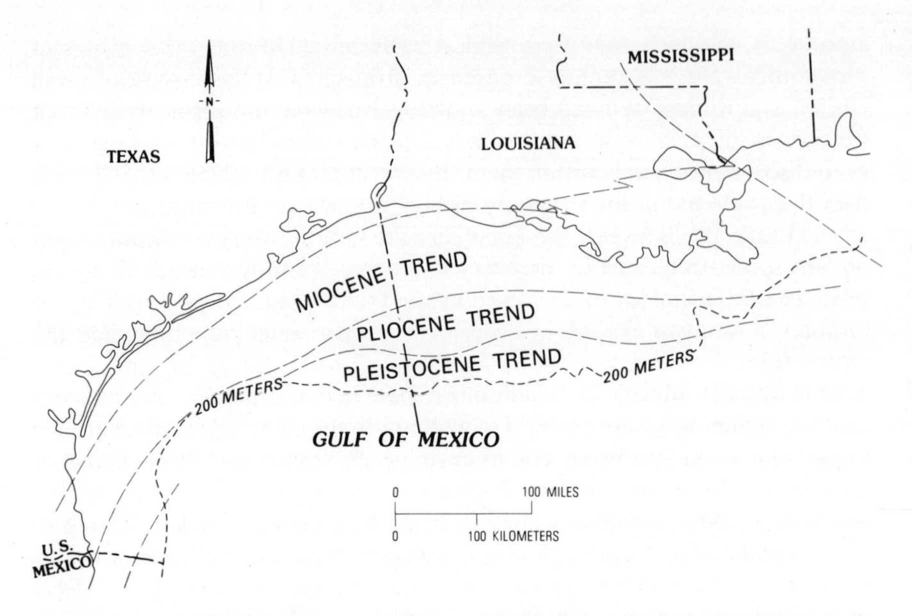

Figure 7.2. The offshore geologic trend areas in our study area in the Gulf of Mexico (from an unpublished map of the Gulf of Mexico prepared by the then Conservation Division of the U.S. Geological Survey for federal Outer Continental Shelf lease sale 38 held in 1975).

missing exploratory wells that were not recorded in the first two sources. Only minor information from this source was added.

4. Initially, wildcat discovery wells, unless specifically identified, were designated as the last wildcat wells drilled before the drilling of the first exploratory or development wells. These designations were revised if the individual field studies published by such sources as the Lafayette Geological Society cited a different wildcat well as the discovery well.

5. The volume of producible oil and gas discovered by each successful wildcat well drilled on federal acreage was supplied by the then Conservation Division, and such estimates for the state-owned acreage were supplied by the Dallas field office of the EIA.

6. In October 1979, Walter Bawiec and I checked the overall completeness and correctness of the data file by comparing the file with the raw data sources at the Metairie offices of the then Conservation Division. From the results of this field check, which was based on a random sample of 250 OCS leases, we were able to state that the file was 95-percent complete and correct.

When I made the request to be allowed access to the files in the Metairie office, I knew there was a chance that we would not be allowed on the premises because we were from a research unit in the Geologic Division and the Metairie office was a field office in the Conservation Division. After several discussions with the Headquarters folks in Reston, we were told that a letter would be sent ahead on our behalf. I did not know at the time that it would be signed by Bob Rioux, deputy chief of the Conservation Division. This letter was our ticket to get inside the vault and at the shelves of archived lease data. When we arrived at the Metarie library, we were welcomed by Sonia LaBee, the very imposing lady who was in charge of the archives. It was immediately obvious that we were in her kitchen and that this was a privilege. We also made our appointed acquaintance with Floyd Bryan, who had been assigned the task of answering any questions about well logs, discovery dates, and reserve estimates.

As we began our comparison of the raw records against our random sample, we ran into the problem of inconsistent well names. As we built the file, we resolved the issue by cross-checking spud and completion dates and total depths. We knew, for example, that a well designation of OCS-1 could be the same as OCS-A-1. We had found out early on that a designation such as OCS-1 could usually be taken to mean that the well was expendable because it had been drilled from either a jack-up or semisubmersible rig, whereas the OCS-A-1 designation meant that the well was permanent because it had been drilled from the A platform; this second well also could be a reentry of the first well. We were told that, although it made no sense, the operators made up the names, sometimes using the convention and sometimes not; so we had to check the scout tickets to be sure. It may not seem like much, but this was important advice. It meant that we had not wasted our time working so long on well names. The biggest surprise was the high degree of accuracy of the International Oil Scouts Association lists of discovery wells. Floyd Bryan went to the well-log vault a number of times to ascertain which well discovered a given field. The Metairie office had a list of discovery wells that sometimes disagreed with the calls made by the Scouts. When this occurred, Bryan's list usually had a later well in the drilling sequence of the lease in question. I remember when he showed us the induction-electric survey on one well, he said something like, "Well, we missed this one, and it has more than 15 feet of pay, which is our cutoff for a discovery well. It is ratty pay, but pay all the same." The point always seemed to go in favor of the calls made by the Scouts. As we were finishing up, Floyd paid us the best compliment that an operations man in a field office can give anyone from headquarters and/or a research unit: he asked for a copy of our complete data file. I also took his request as evidence that we had gone about as far as we could go and that it was time to stop collecting data and to start analyzing it.

In the process of checking out our data file in Metairie, we participated in a little of the human drama of the vault room. I remember Sonia LaBee as a tough-minded lady from an incident that involved a geologist who did not return a well log as he had been expected to do. She did a war dance behind the counter and then wondered how this geologist could do geology if he was unable to do simple tasks. Her real threat was expulsion from the vault. To me, it seemed that the geologist had committed the trivial infraction of not returning a well log on time, which was quite excusable according to his explanation. Yet he stood before her like a naughty child awaiting his fate to be decided. He walked away happy after having received only a stern tongue lashing. She ran a tight ship because a misplaced well log was a lost well log. The filing system required *absolute* order.

We returned to Reston satisfied with the results of our field check of the 250 random sample leases. We had a valid set of data and had established an audit trail on the procedure used in its preparation. Proof that a valid data set had been prepared for the Gulf of Mexico was not enough to convince many of our colleagues that we were going to be successful in forecasting the future rates of discovery of oil and gas fields in that region. The consensus was that, because the exploration history in the Gulf of Mexico was so complicated, no amount of analysis would make our data reveal its secrets in any organized fashion.

This allegation was based on the belief that discovery process modeling was a new field that had only been tested on a few onshore areas. The arguments that I usually heard went as follows: Because these few already-examined onshore areas may have been the only areas where exploration had been unrestrained and where larger fields had been found earlier than the smaller fields, some order had been imposed on the discovery process, which, in turn, could be used as a basis for forecasting future rates of discovery.

It would be unfair to say that attitudes about the usefulness of discovery process modeling in making predictions as to the amount of oil and gas remaining to be discovered and future rates of discovery were not starting to change. Although the idea that there might be some order to the discovery process was beginning to grow, we were given the impression that we had gone too far when we proposed using it in the Gulf of Mexico study. We were told repeatedly that because water depth was such an overriding factor in the decision-making process, it alone would determine any systematic variation that we might find. If this were not enough to squelch our desire to model the exploration history of the Gulf of Mexico, we were told that we would have to stop when we realized that a lot acreage had been held off the exploration market for long periods while state and federal governments argued in court over who was to own which parcels of land. The changeover from exploring for crude oil to exploring for natural gas also was cited as an additional factor that would prevent any discovery process model from working correctly.

In the autumn of 1979, this ill humor overhung our effort to apply discovery process modeling to the Gulf of Mexico. True enough, we did not have any experience in offshore regions, but it was worth the effort to see how applicable these models were in regions where the constraints of water depth, lease sales, and jurisdictional disputes were believed to have largely controlled past exploration history. It seemed to me that data-blocking techniques could be used to create data subsets in which the geologic trends were made approximately colinear with water depth. This would help to sort out the drilling and the discovery data into more homogeneous groups, thereby, at least partially, dealing with the issue that exploration tended to move into deeper and deeper water over time. The changeover from exploring for crude oil toward exploring for natural gas was accommodated easily by performing the analysis in barrels of oil equivalent. The problems introduced into the historical discovery rate data by the sequestering of acreage during the state versus federal ownership disputes had to be fairly limited because only a small proportion of the total offshore acreage was involved.

We were by no means starting out on a frontier adventure in the analysis of oil and gas exploration data. By 1979, a fairly well developed standard set of discovery rate profiles, which we believed characterized the behavior of exploration for the unrestricted case, was available for oil and gas exploration in onshore regions. Examples ranged from the small scale of a sedimentary basin up to the entire onshore United States (Hubbert, 1967, 1974; Ryan, 1973a, b; Root and Drew, 1979, 1981). We had faced the situation of forecasting future rates of discovery in the Denver basin, where a large block of acreage had been sequestered by railroad ownership (Drew et al., 1978, 1980), and we had prepared and analyzed discovery rate data for many basins in the United States, which were subsequently published as an atlas of discovery rate profiles (Drew et al., 1983).

The data obtained from these studies resulted in a body of knowledge that we would use to determine how the different methods used by the oil industry to explore for oil and gas in the Gulf of Mexico and in onshore regions varied. My suspicion was that the pattern of historical discovery of oil and gas in this offshore region would not be very different from that which had occurred onshore; for example, it seemed reasonable that the L-shaped, or step function, pattern of the aggregate discovery rate profile, which had been established for the entire United States (Figs. 3.3 and 3.7) and was typical for areas in which small amounts of petroleum had been discovered, such as the Denver basin (Fig. 4.21), would be typical of the past rates of discovery in the Gulf of Mexico. It also seemed reasonable to assume that the oil companies would seek out the best prospects that had the biggest fields as early as possible in the exploration history no matter how great or ever present the physical and political restraints. They had strong marine engineering staffs to solve the technical problems they faced in offshore exploration and production. They also had capable legal and public relations departments to pursue their interests in the political arenas.

Several tasks were performed on the Gulf of Mexico data. The first of these was the determination of whether the aggregate discovery rate profile was L-shaped. The reference profile that we commonly used in this type of comparison was for the Permian basin (Fig. 7.3). The second operation performed on these data was an analysis of the systematic composition of an aggregate discovery rate profile by individual and average field sizes as a function of cumulative footage and the number of wildcat wells. Reference graphs of these data for the Permian basin are shown in Figs. 3.9 and 3.10. The behavior of the individual field size discovery rate profiles determines whether a discovery pro-

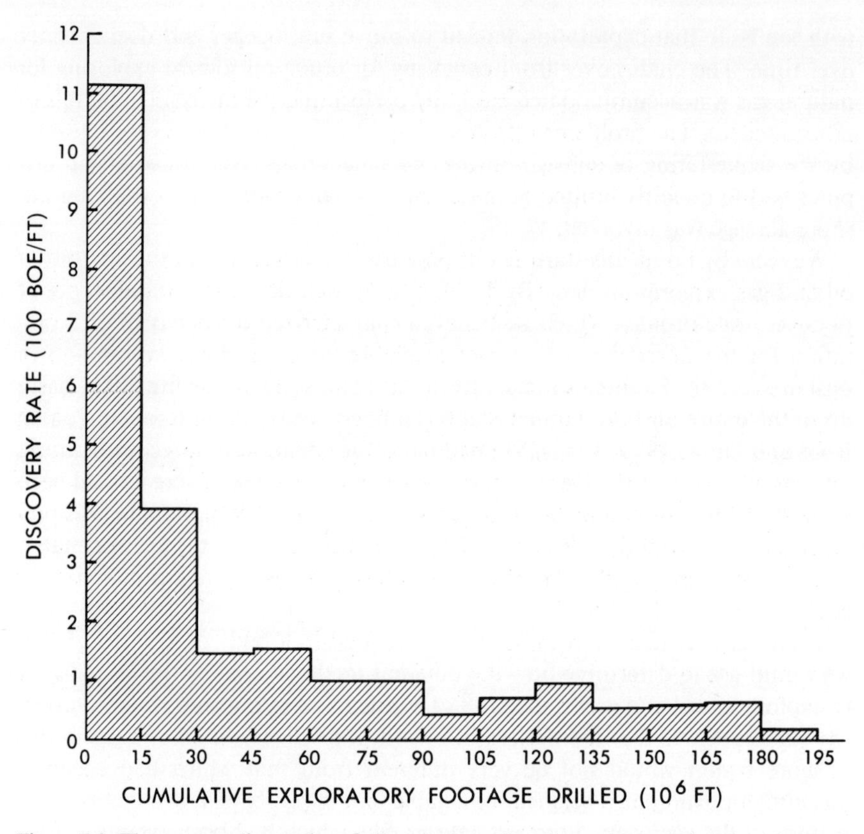

Figure 7.3. The average discoveries of oil and gas per foot for each 15 million feet of exploratory drilling in the Permian basin, 1921—74 barrels of oil equivalent = 1 barrel of crude oil or 5270 cubic feet of wet gas (Root and Drew, 1981).

cess model can be used effectively to produce an assessment of the undiscovered oil and gas resources and a forecast of the future rates of discovery within a region.

The historical discovery data must display diminishing rates of return to at least several of the largest field size classes before a discovery process can be effectively used. The discovery rate data displayed in Fig. 3.9 show that the required diminishing rates of return to wildcat drilling had occurred before 1945. By that date, one-half of the total oil and gas discovered through 1974 had been discovered in only 4000 of the 30,417 exploratory wells drilled through 1974. Most of the oil and gas discovered as a result of drilling these 4000 wells was in fields that contained more than 100 million barrels of oil equivalent each. As exploration proceeded in the basin, the contribution to the total discoveries of oil and gas from the various field size classes progressively shifted away from the largest classes toward the smaller field size classes. This type of shift had been established as a characteristic of the progressive maturity of returns to exploratory drilling in onshore exploration plays and basins within the United States before we began the analysis of the Gulf of Mexico discovery rate data. This progressive shifting toward smaller and smaller discoveries is the diminishing rate of return behavior that a discovery process modeler uses to make a forecast of future discovery rates.

We had also encountered regions, such as the Michigan basin, where multiple exploration plays had unfolded in a temporally disjoint manner and other regions where the discovery rate profiles were a composite of multiple plays that occurred simultaneously. We knew that several plays had occurred in the Gulf of Mexico and expected that we would encounter patterns in the discovery rate profiles that were caused by either of these types of compositions. Figures 7.4–7.8 display the behavior of discovery rates in the Michigan basin where three temporally disjoint plays unfolded. The three spikes in the incremental discovery rate profiles mark the start of the three exploration plays (Figs. 7.4 and 7.5). The contribution made to the graph of cumulative discoveries by the

Figure 7.4. Incremental discovery rate profile for the 160 fields in the Michigan basin (barrels of oil equivalent per foot versus cumulative footage drilled) (Drew et al., 1983).

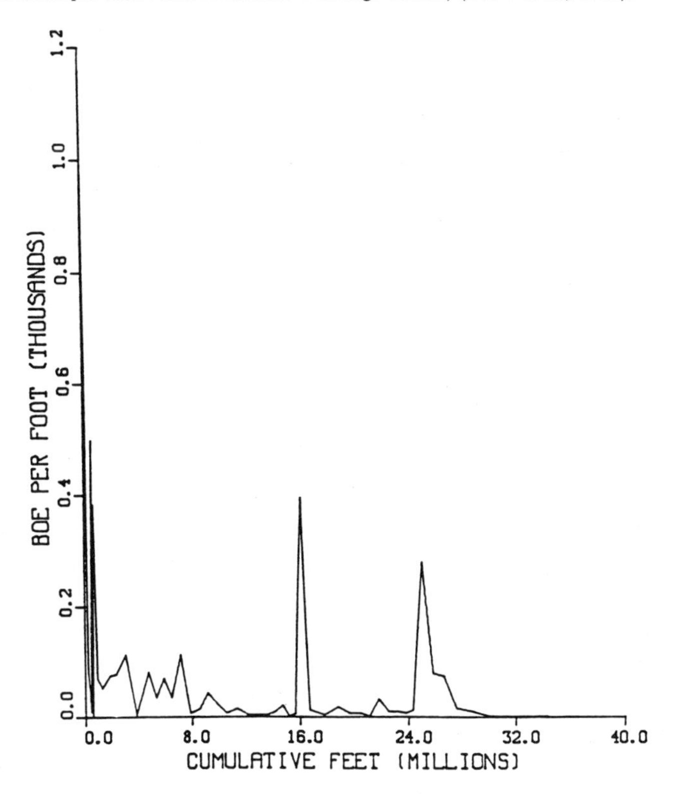

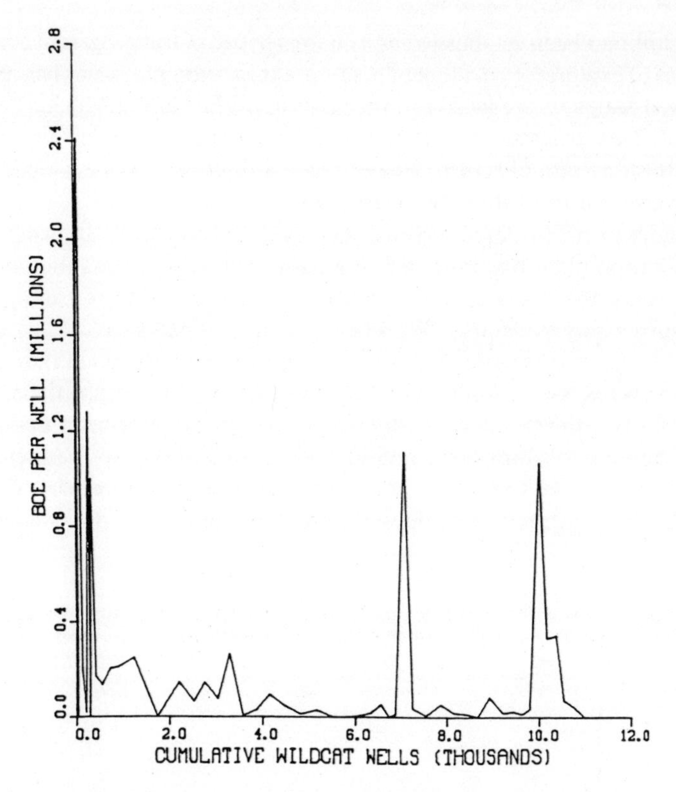

Figure 7.5. Incremental discovery rate profile for the 160 fields in the Michigan basin (barrels of oil equivalent per well versus cumulative wells drilled) (Drew et al., 1983).

initiation of these three plays is displayed as the marked three jumps in Fig. 7.6.

In the Michigan basin, we found that the discovery rate of the smaller fields was approximately linear (Fig. 7.7), whereas the discovery rate of the larger fields was disjoint (Fig. 7.8). This pattern of nearly linear and continuous discovery rate behavior for the smaller field size classes is typical of most regions where multiple exploration plays have unfolded in a temporally disjoint manner. In those basins where *many* exploration plays have unfolded, some simultaneously and some disjoint, this identical behavior also occurs; the Permian basin is the archetypal example. In this basin, the discovery rate profiles are linear or nearly linear for the smaller field classes.

As we started to analyze the discovery rate data from the Gulf of Mexico, I felt that we had collected and analyzed enough data to provide us with information about most of the possible configurations that could be found in oil and gas discovery rate profiles. The question, then, was how could the exploration history of the Gulf of Mexico be so confounded by physical, legal, and institutional restrictions that an entirely new type of discovery rate profile had to be developed and that we might not get a meaningful diminishing rate of return relation to use to predict the number of fields remaining to be discovered and to forecast their future rates of discovery? We had been given a lot of advice that implied that that was what we were going to find.

The aggregate discovery rate profile for the entire history of exploration in the Gulf of Mexico is displayed in Fig. 7.9. This aggregate profile shows that early in the exploration history of this region (through 1958), the rate of discovery was relatively high, remaining above 7 million barrels of oil equivalent per well. From 1959 onward, the discovery rate varied around 2 million barrels of oil equivalent per well. The high rate of discovery recorded during the early years of exploration in the Gulf of Mexico is, in large measure, the consequence of the discovery of several giant oil and gas fields, such as Bay Marchand Block 2 and the combined South Pass Block 24–27 fields. These fields were included in the data set even though they are not entirely offshore; the normal convention was followed here in calling these offshore fields.

Figure 7.6. Cumulative oil and gas discoveries versus cumulative wells drilled for the 160 fields in the Michigan basin (Drew et al., 1983).

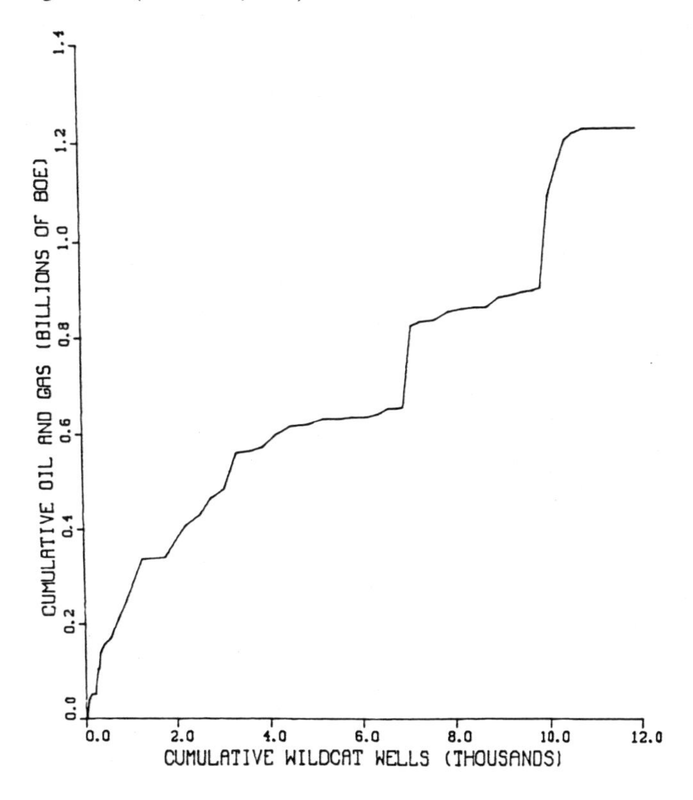

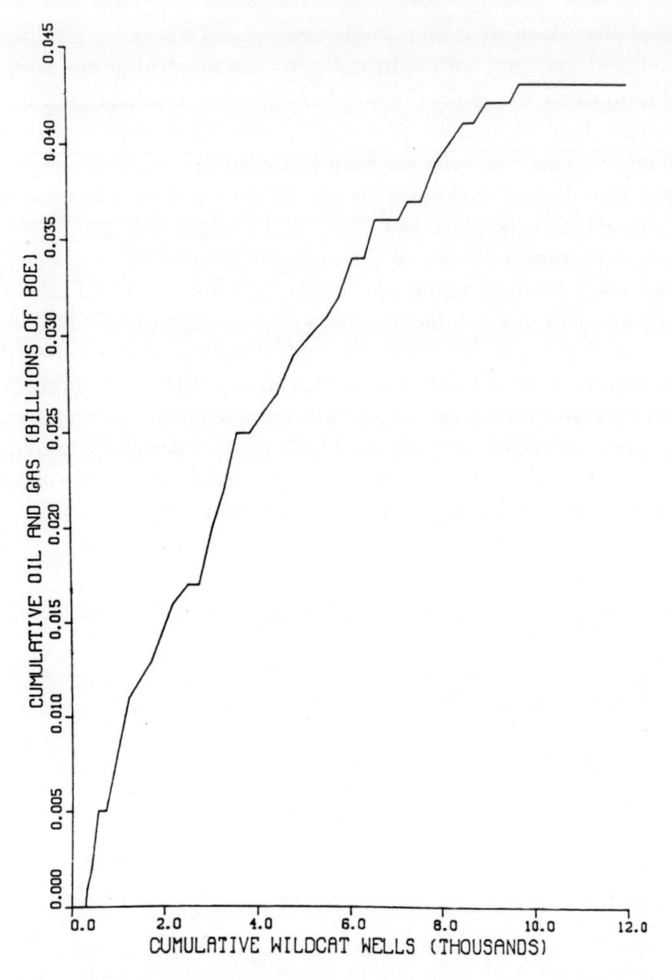

Figure 7.7. Cumulative oil and gas versus cumulative wells for the 43 fields in the size range 0.8 million–1.5 million barrels of oil equivalent in the Michigan basin (Drew et al., 1983).

So, the L-shaped character of the aggregate discovery rate profile was a fortuitous consequence of having included, for the sake of completeness, the very large discoveries made in a small near-shore subregion of the Gulf of Mexico. Although it is impossible to say that no discoveries of this size will ever be made again anywhere in the study area, it is reasonable to say that by using data from the late 1970s, this is a most unlikely expectation. For the purpose of our investigation, it was much more important to determine if a regular pattern of diminishing returns to wildcat drilling existed in the 1959–1978 data in a meaningful set of subregions so that a discovery process model could be used to estimate the number of fields by size class remaining to be discovered and their rate of discovery in the future. During this 20-year period, the drilling of nearly 6000 exploratory wells in the study area resulted in the discovery of approximately 400 oil and gas fields.

During the 1959–1978 period, the aggregate discovery rate oscillated around 2 million barrels of oil equivalent per well. Initially, this constant rate of return was identified as being similar to what had occurred in the Permian basin where the aggregate discovery rate profile was held up by discovering larger fields at increasingly deeper depth intervals over time than would have otherwise been made had the exploration effort been confined to the shallower depth

Figure 7.8. Cumulative oil and gas versus cumulative wells for the 13 fields in the size range 12.1 million–24.3 million barrels of oil equivalent in the Michigan basin (Drew et al., 1983).

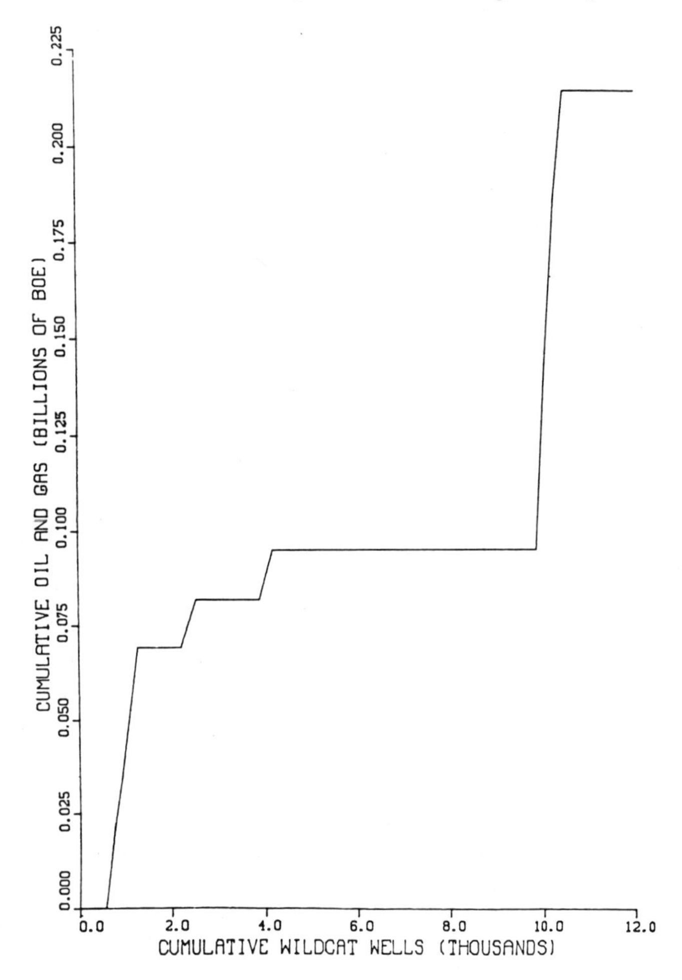

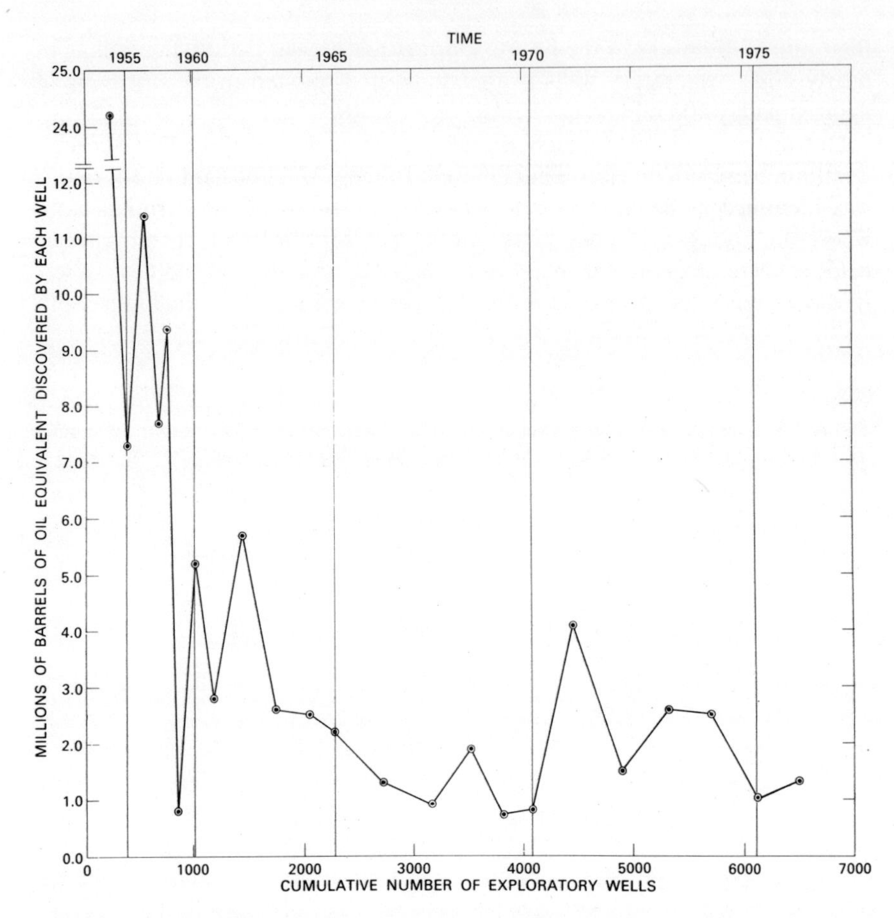

Figure 7.9. Rate of discovery of oil and gas in the Gulf of Mexico study area in state and federal waters 0–200 m deep.

intervals. In the Gulf of Mexico, the equivalent type of discoveries had to be the result of the seaward migration of the exploration effort.

The vast majority of the oil and gas discovered in the Gulf of Mexico between 1959 and 1978 was found, on average, in 32 lease sales to the oil industry by the federal government. One of the first graphs we produced in our analysis is displayed in Fig. 7.10. This graph shows the relation between the amount of crude oil and natural gas discovered on the federal acreage as a function of the cumulative number of leases drilled. It also reveals that the volume of producible crude oil and natural gas discovered on the federal acreage, on a lease-by-lease basis, has been nearly constant across all lease sales. The curvature at the end of this graph, which covers the last four or five lease sales, suggests that the rate of return to wildcat drilling started to decrease after drilling on about 1300 leases. This curvature was attributed to the incomplete record-

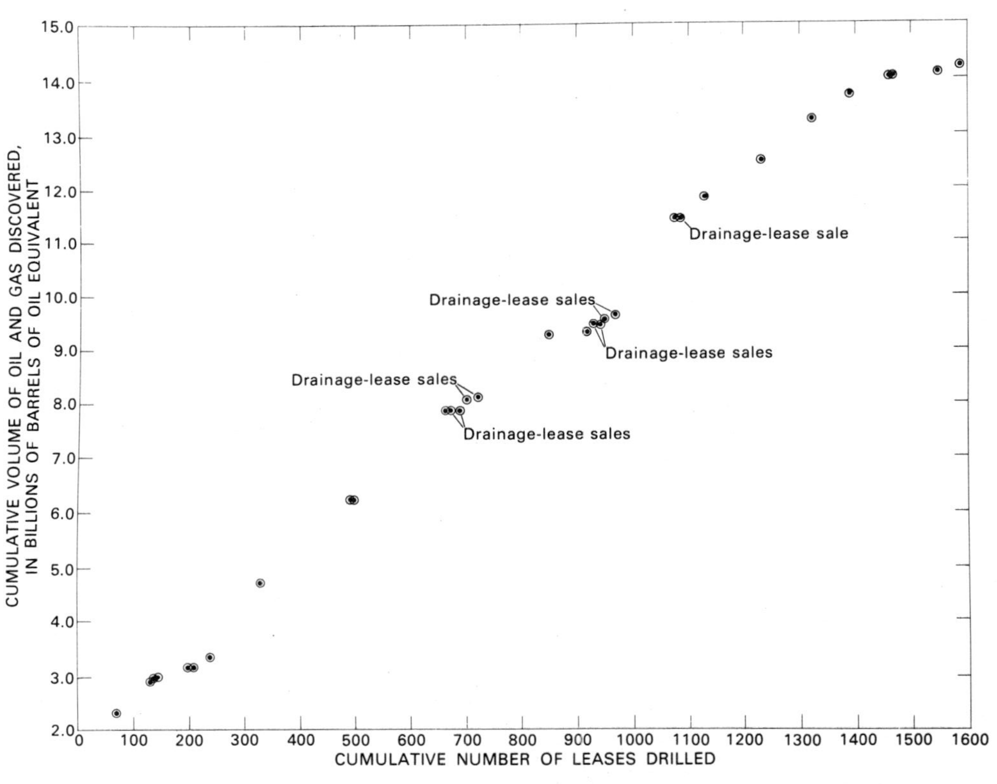

Figure 7.10. Cumulative volume of oil and gas discovered on federal Outer Continental Shelf acreage in the Gulf of Mexico study area on which leases were sold by February 4, 1975—volume reported as barrels of oil equivalent (BOE) per cumulative number of leases. The first data point represents the 2.3 billion BOE discovered on acreage that is now under federal jurisdiction but that originally was leased by the state of Louisiana. All this oil and gas was discovered on leases sold before the first federal sale, which was held October 13, 1954. The last data point includes drilling done through December 31, 1978, on acreage on which leases were sold by February 4, 1975.

ing of the growth to the reserve estimates for the fields discovered on the acreage leased in the last few lease sales. Although we made this assumption, we wondered if it could be the first manifestation of diminishing returns in the aggregate discovery rate data for the region as a whole. The thought did occur to us that this could mean that the aggregate discovery rate was beginning to fall through a transition phase to a new, lower plateau. My suspicion at the time was that the curvature at the end of this graph was an artifact of the incomplete reporting of the size of oil and gas reserves that had been discovered. Therefore, I believed that the constant rate of aggregate discovery of oil and gas at about 2 million barrels of oil equivalent per wildcat well would continue for a while longer.

The "tight" collections of data points in the middle sector of Fig. 7.10 represent the results of drilling a few exploration leases included in nine drainage-lease sales held between 1962 and 1971. The general intent of a drainage-lease sale is to sell leases on structures previously proven to be productive. In most drainage sales, however, a few exploration leases are included. The data from the drainage-lease sales plot as tight clusters because not more than three new field discoveries were made on the acreage leased in any of these sales.

What does Fig. 7.10 really tell us? Does it not, at least, show that the collective strategy of the oil and gas industry worked effectively against the high-cost exploration constraints associated with moving into continually deeper and deeper water depths in the Gulf of Mexico, which caused the rate of discovery to be held up to a near constant level? In general, is it not true that the only way to keep the rate of discovery from exhibiting diminishing rates of return is to continue to enter new and untested volumes of rock where fields are discovered that are larger than would be discovered in the rock volumes previously tested? This does not imply that the aggregate discovery rate profile has to have any particular analytical form. In this regard, I remember thinking that it was curious that the movement of exploration into deeper and deeper water over time, along with the changeover from discovering crude oil to discovering natural gas and with other institutional changes, had been so systematic as to result in a constant rate of return.

The companion graph (Fig. 7.11) was produced by plotting the cumulative volume of oil and gas discovered in this region against the cumulative number of wildcat wells, showing essentially the same pattern of aggregate discoveries. The graph in Fig. 7.11, however, does change its slope somewhat after the drilling of about the 1000th wildcat well (in 1966). This difference in slope shows that the rate of oil and gas discovery per wildcat well, on a barrels of oil equivalent basis, was somewhat higher before 1966 than after. When the first 1000 wildcat wells were drilled, exploratory drilling was confined, in large part, to Miocene and Pliocene rocks on the federal OCS acreage off Louisiana. The second segment of the graph covers the 1967–1978 period when exploration was focused on exploring for fields in the gas-prone Pleistocene rocks located, for the the most part, in the West Cameron (offshore Louisiana) and High

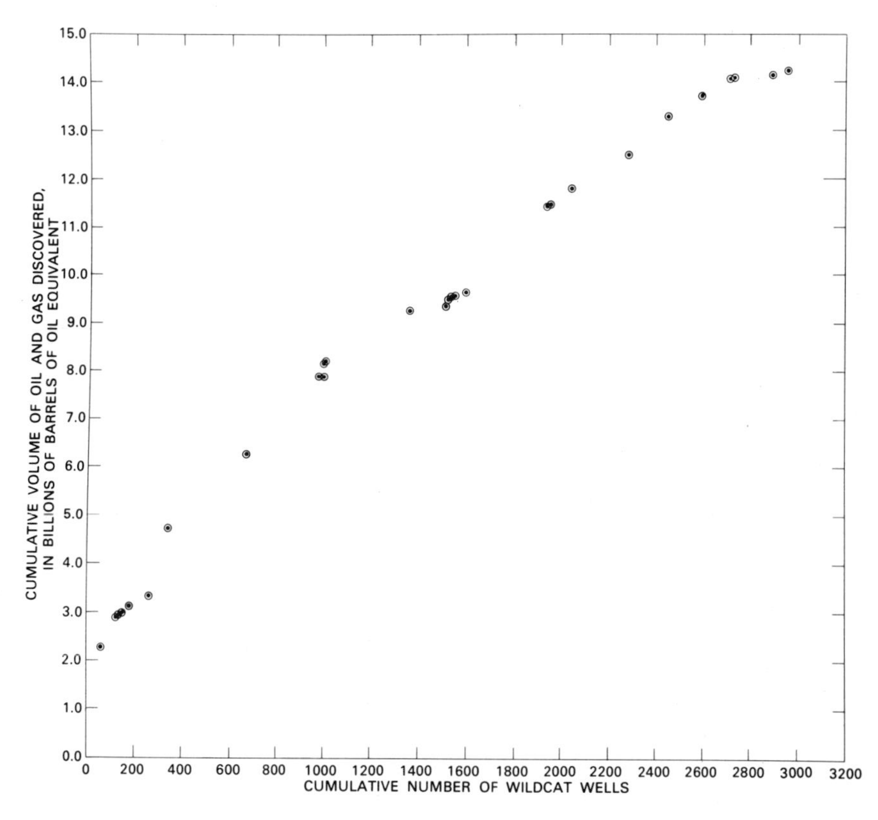

Figure 7.11. Cumulative volume of oil and gas discovered on federal Outer Continental Shelf acreage in the Gulf of Mexico study area on which leases were sold by February 4, 1975—volume reported as barrels of oil equivalent (BOE) per cumulative number of wildcat wells. The first data point represents the 2.3 billion BOE discovered on acreage that is now under federal jurisdiction but that originally was leased by the state of Louisiana. All this oil and gas was discovered on leases sold before the first federal sale, which was held October 13, 1954. The last data point includes drilling done through December 31, 1978, on acreage on which leases were sold by February 4, 1975.

Island (offshore Texas) subareas. This Pleistocene play has produced mainly natural gas, whereas the play in the Miocene and the Pliocene rocks produced a much larger proportion of crude oil.

The change in the slope of the cumulative discovery profile after 1000 wildcat wells reveals the shift of exploration effort away from crude oil and toward natural gas (Fig. 7.11). The slope change exists because the gas fields discovered in the later segment of exploration are somewhat smaller on a barrels per oil equivalent basis than the oil fields discovered earlier. Being able to see such a transition in the data was very welcome. It meant that perhaps the data could be partitioned into meaningful subsets that would be useful for forecasting. The similarity in form of the two graphs shown in Figs. 7.10 and 7.11 reveals that

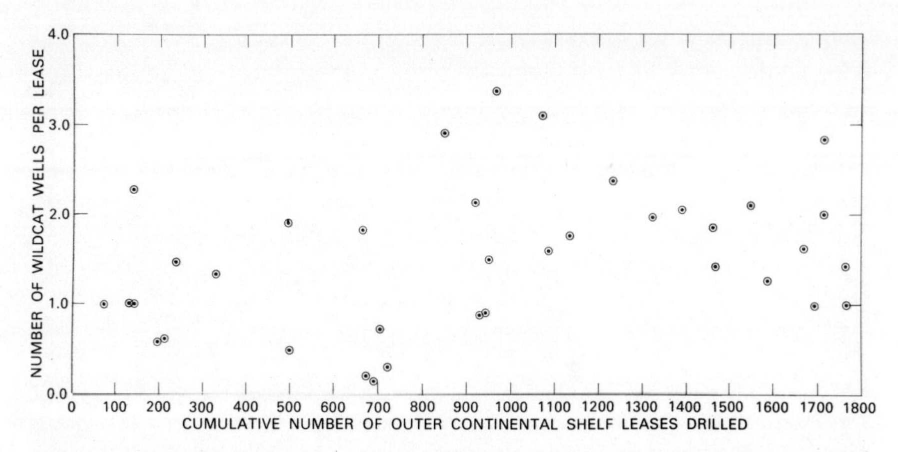

Figure 7.12. Relation between the number of wildcat wells drilled per lease and the cumulative number of leases drilled in the Gulf of Mexico study area. The figure reflects drilling done through December 31, 1979, on acreage on which leases were sold by June 29, 1979 (sale 48).

the offshore operators have been consistent through time in the manner in which they have evaluated leases. The data displayed in Fig. 7.12 shows that, on a wildcat well per lease basis, about two wildcat wells were used, on average, to evaluate a lease across the history of exploration on federal acreage through 1979. This conclusion may not seen worthy of mention, but it was an important bit of news. At the time, we had been told repeatedly that we could not forecast the future rates of discovery in the Gulf of Mexico by using wildcat wells to drive a discovery process model. The point was that an offshore well was not like an onshore well because these wells were drilled on small leases at great costs. In addition, the geology of a rollover gas trap (a common exploration target in Pleistocene rocks) was very much simpler than the geology of a salt dome (a common exploration target in the Miocene rocks). This suggests that more wildcat wells would have to be drilled to evaluate the latter type of trap. However, the data in Fig. 7.12 show that the information used to make the decision about whether to keep or drop a lease was gained from drilling about the same number of wildcat wells across the entire history of exploration on federal acreage in the Gulf of Mexico. It is my recollection that most of this type of analysis provides new information on how consistently the exploration operators had performed the task of exploring for oil and gas in the offshore areas. To me, the work was necessary to show why a discovery process model could be used to estimate the number of fields remaining to be discovered by size class and to forecast their rate of discovery in the future or why it could not be done. As the pattern of consistent behavior in the Gulf of Mexico was built up from the data analysis, I could find no reason not to proceed.

The next step was to determine if the discovery rate profiles for the individual field size classes contained enough of a diminishing rate of return effect to allow us to calibrate the Arps and Roberts discovery process model. I wanted to use a less clumsy criterion to fit the model to the data than the aggregate criterion we used to calibrate the discovery process model in the Permian basin. What we wished to do was solve for the efficiency of discovery and the ultimate number of fields simultaneously within each field size class. This approach could easily put too many demands on the data. Before we attempted to fit the model to the individual field size discovery rate profiles, I plotted all the data by hand.

I believe in plotting all the data that I am going to use in any model by hand on graph paper. It may be a character flaw in this era of computer graphics packages not to use them. To me, a great deal is lost when I do not take the hours or days to plot each point by hand. It is the knowing of the data that is lost when the plots are printed out on a machine. I want to use the dictum John Griffiths inculcated in his students—plot and know your data. His warning to me as student was not to let the null hypothesis put meaning into my data. It is always easy to forget that all models have assumptions about the behavior of the data used to fit them. Griff would say that parametric statistical analysis was his business, and he made it clear that it began with graph paper and a pencil. When he taught that the arrangement of the data points in bivariate plots are shadows of the arrangement of data in some higher dimensions, his lessons on covariation came alive in the marking of the graph paper and the connecting of the points. In the plotting of the Gulf of Mexico data, I wondered how discoveries that were made in one field size class were connected in the higher dimensions of reality beyond my graph paper. How would we connect the activity of wildcat drilling and discovery in one field size class to what had gone on in all the other field size classes? In fitting the Arps and Roberts model, the assumption was made that no damage was done by considering them to be independent.

The first set of graphs that were constructed are displayed in Figs. 7.13–7.16. Examination of the graphs shows why the rate of discovery of oil and gas on an oil equivalent basis oscillated around 2 million barrels of oil equivalent per exploratory well on the federal acreage from 1958 through 1978 (Fig. 7.9). In Fig. 7.13, note first that all but one of the very large fields (size classes 18 and 19) were discovered by the time 500 wildcat wells had been drilled in the study area. The end points for each size class in barrels of oil equivalent are shown in Table 6.2; for example, field size class 19 ranges in size from 777.2 million to 1554.4 million barrels of oil equivalent.

The discovery of all but one of these large fields had been made by 1958, when about 800 total exploratory wells had been drilled (Fig. 7.9); during this same time, 500 wildcat wells had been completed. We will switch to exploratory wells from wildcat wells only as required when discussing the data displayed in Fig. 7.9. My own preference is to use only wildcat wells in such analyses, but I recognize that it is not always possible to do so. The number of exploratory wells is a common basis used to plot aggregate discovery rate data. Exploratory wells include wildcat wells and other classes of wells usually

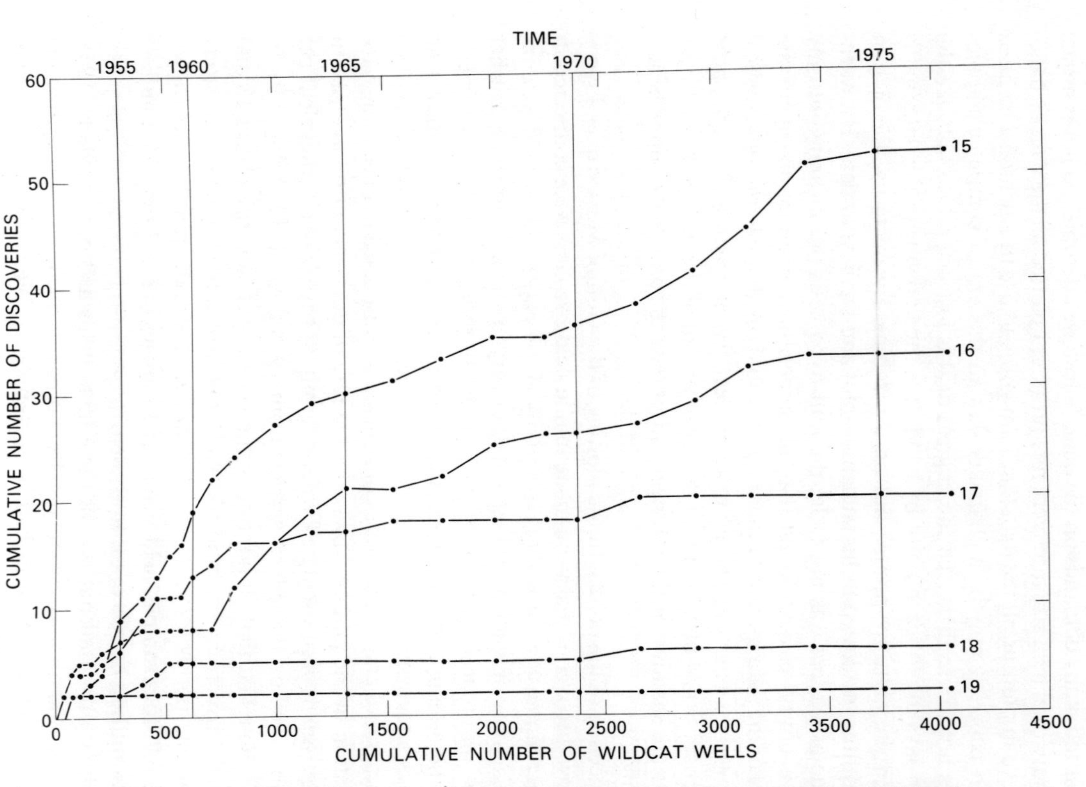

Figure 7.13. Cumulative number of oil and gas fields discovered by December 31, 1976, in size classes 15–19 in combined state and federal waters 0–200 m deep in the Gulf of Mexico study area.

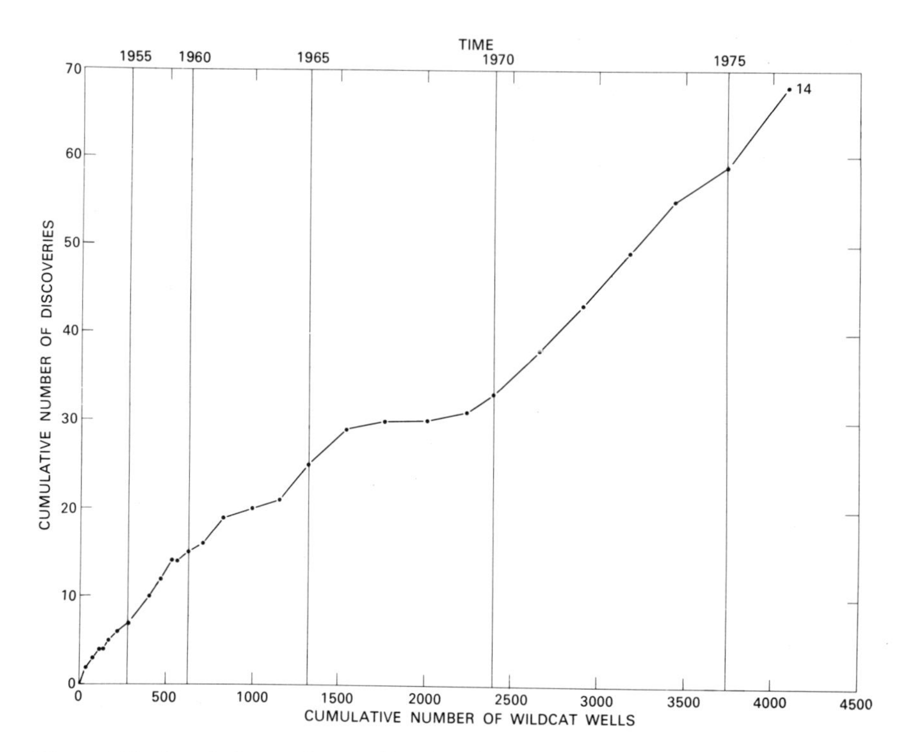

Figure 7.14. Cumulative number of oil and gas fields discovered by December 31, 1976, in size class 14 in combined state and federal waters 0–200 m deep in the Gulf of Mexico study area.

intended to find new pools within a previously discovered field. Although both bases were used in the Gulf of Mexico study, the use of exploratory wells was limited to the aggregate discovery rate graph.

The data in Fig. 7.13 show that the history of discovery of oil and gas in the Gulf of Mexico for the combined federal and state waters exhibits the commonly observed attribute of the early discovery of the largest fields. These included two giant fields found in the tidewater region off Louisiana mentioned above (Bay Marchand Block 2 and the combined South Pass Block 24–27 fields). Even with the removal of these two fields from the data set, the early discovery of all but one of the remaining fields in size classes 18 and 19 suggests that the exploration history of the Gulf of Mexico, at least with regard to the discovery of the largest fields, is similar to the history of exploration in most other petroliferous areas of the world.

The systematic pattern of discovery in the Gulf of Mexico extends far beyond the early discovery of the largest fields. There is a systematic shifting of the discovery rate profiles from field size class 19 down through field size class 15 across the drilling of nearly 2300 wildcat wells or through 1969 (Fig. 7.13). This

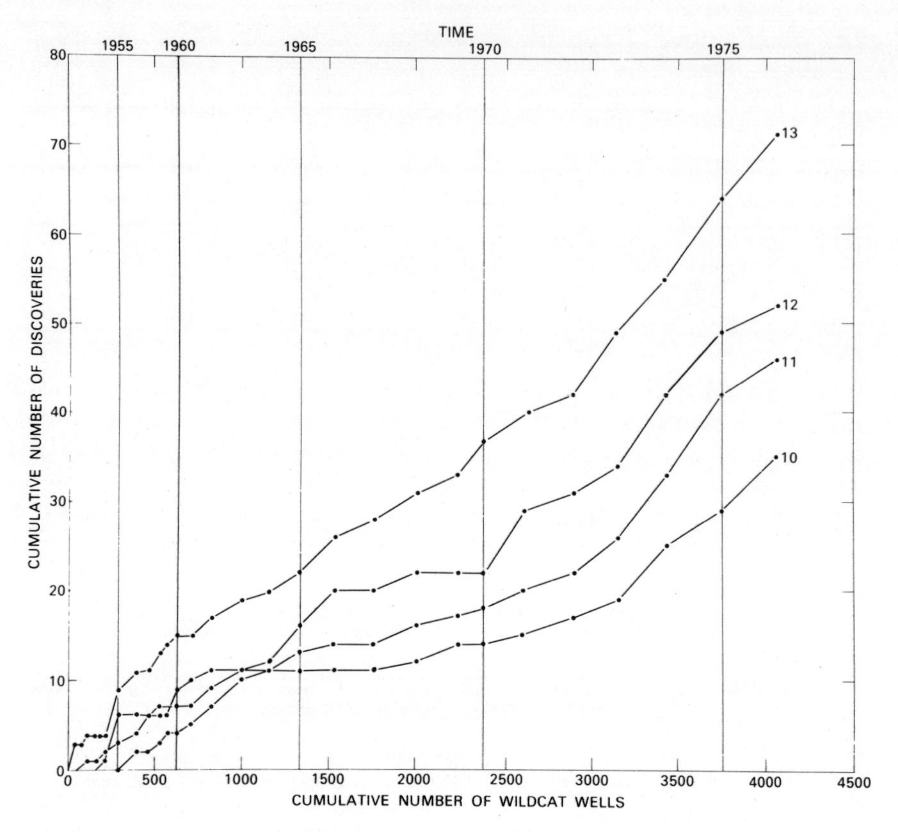

Figure 7.15. Cumulative number of oil and gas fields discovered by December 31, 1976, in size classes 10–13 in combined state and federal waters 0–200 m deep in the Gulf of Mexico study area.

progression can only mean that, from the 1940s through about 1969, the exploration operators were able to selectively choose the better acreage earlier in the exploration history of the Gulf of Mexico. The pattern exhibited in the profiles shown in Fig. 7.13 (through 1969 and the drilling of 2300 wildcat wells) is very similar to those observed in onshore regions where the exploration process has been unencumbered by legal and institutional restraints.

Around 1970, this pattern of diminishing returns to wildcat drilling changed as the slopes of the cumulative discovery rate curves of four of the five largest field size classes broke upward (see Fig. 7.13, segment between 2500 and 4100 cumulative wildcat wells). The same upward break in slope occurs in the next five smaller field size classes (Figs. 7.14 and 7.15). This change in the cumulative discovery rate profiles was caused by the shift in emphasis from exploration and discovery of mainly crude oil to natural gas. The search for natural gas was principally focused in the gas-prone Pleistocene rocks, which occur in the East and West Cameron, Louisiana, subareas and in the High Island subareas of Texas.

This upward break in the individual field size discovery rate profiles caused the aggregate discovery rate profile (Fig. 7.9) to rise in 1971 and to maintain itself at about 2 million barrels of oil equivalent per exploratory well. While being good news for producers and consumers of energy, these upward breaks are sometimes bad news for the discovery process modelers because the model will not produce meaningful results unless a partitioning of the data can be made where diminishing rates of return to wildcat drilling exist. Although these breaks did not prevent us from proceeding, they did make the task of forecasting a little more complicated. We were able to partition the data according to geologic trends into two subsets that were geologically and statistically meaningful. The large growth fault that exists between Pliocene and Pleistocene trends was used to separate the lease blocks into the two subareas (Fig. 7.2). The resulting subareas are the Miocene-Pliocene and the Pleistocene trends. The map used for this purpose is an unpublished map of the Gulf of Mexico prepared by the then Conservation Division in Metairie for federal OCS lease sale 38 that was held in 1975 (Fig. 7.2).

Choosing a scheme to block a set of data after the data are collected is never

Figure 7.16. Cumulative number of oil and gas fields discovered by December 31, 1976, in size classes 3–9 in combined state and federal waters 0–200 m deep in the Gulf of Mexico study area.

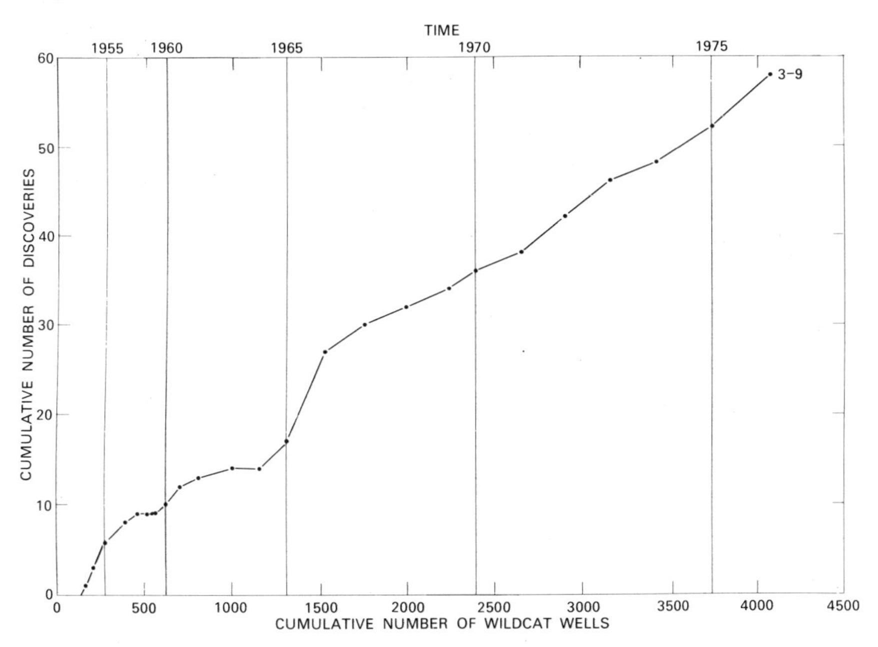

a totally satisfying activity. There is always the danger of knowing beforehand what you specifically need from a data analysis and then making sure you get it by peeking at the data before it is blocked. Because I am a conservative model builder by nature, I tend to err toward using very simple data-blocking schemes that can be defended on first principles or, at least, by the law of parsimony. I did not want to have to defend a blocking scheme whose purpose was to isolate the acreage in the vicinity of the Pleistocene depocenter. I could imagine that maybe as many as one-half of the lease blocks that I or anyone else selected to be inside this area could be quibbled over. Instead, I chose to use the large Pliocene-Pleistocene growth fault. Nobody could pick at that decision; at least, no geologist would. The locations of this fault are well known and could be used easily to partition the acreage where much of the exploration for natural gas occurred after 1970.

The graphs of the cumulative number of discoveries made in each trend by size class for the Miocene-Pliocene trend are shown in Figs. 7.17–7.19, and corresponding graphs for the Pleistocene trend region are shown in Figs. 7.20 and 7.21. Inspection of the cumulative discovery graphs shown in Fig. 7.17 for the Miocene-Pliocene trend versus the corresponding graphs shown in Fig. 7.13

Figure 7.17. Cumulative number of oil and gas fields discovered in size classes 15–19 in combined state and federal waters 0–200 m deep in the Miocene-Pliocene trend of the study area in the Gulf of Mexico.

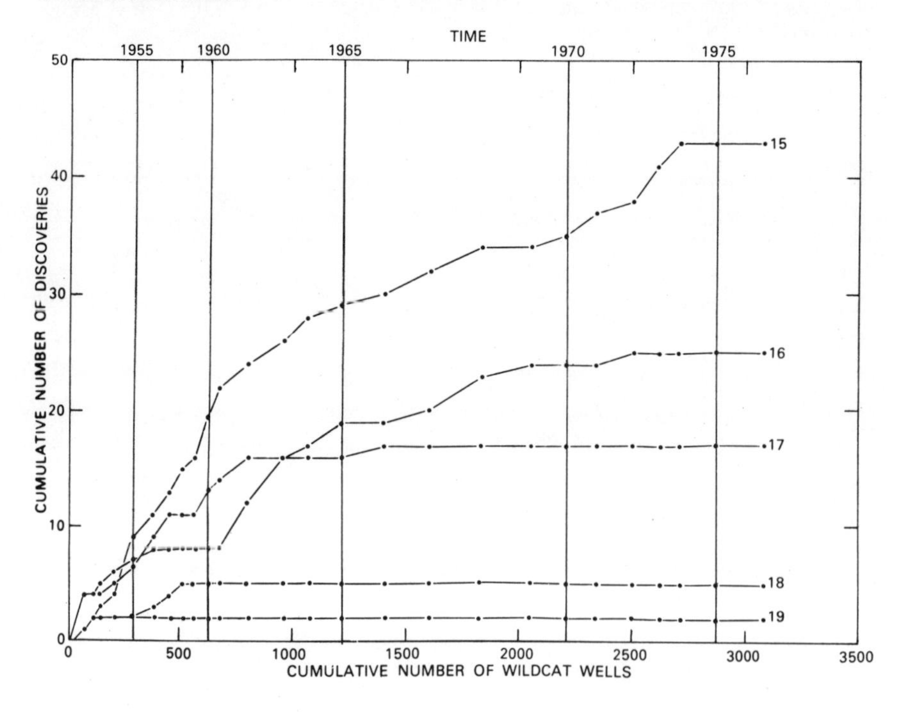

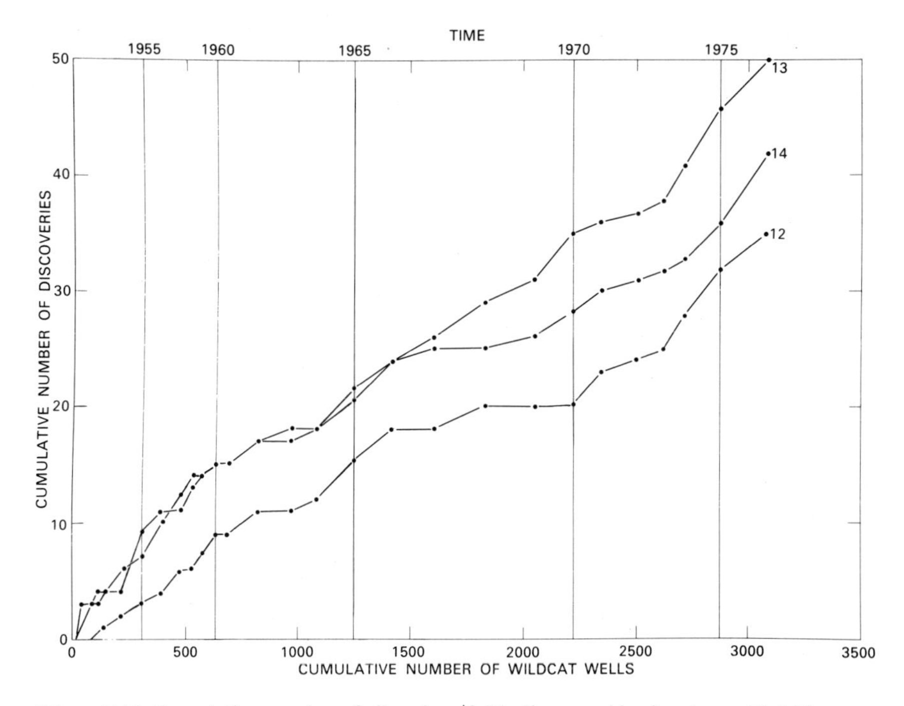

Figure 7.18. Cumulative number of oil and gas fields discovered in size classes 12–14 in combined state and federal waters 0–200 m deep in the Miocene-Pliocene trend of the study area in the Gulf of Mexico.

Figure 7.19. Cumulative number of oil and gas fields discovered in size classes 9–11 in combined state and federal waters 0–200 m deep in the Miocene-Pliocene trend of the study area in the Gulf of Mexico.

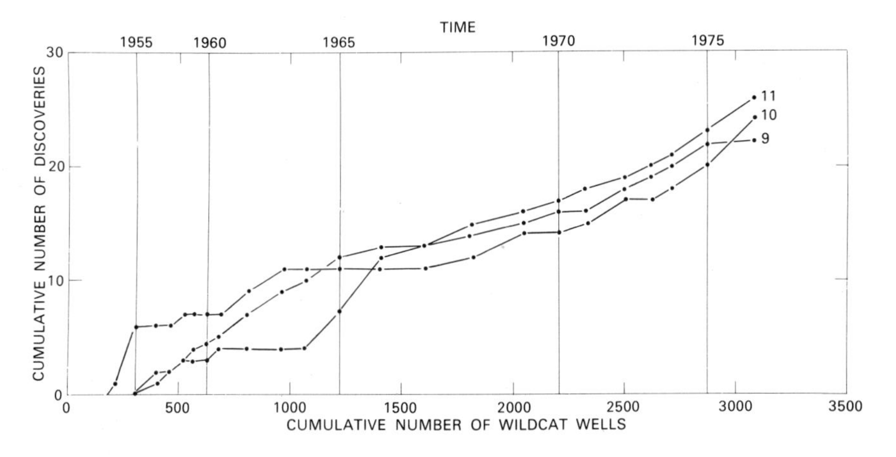

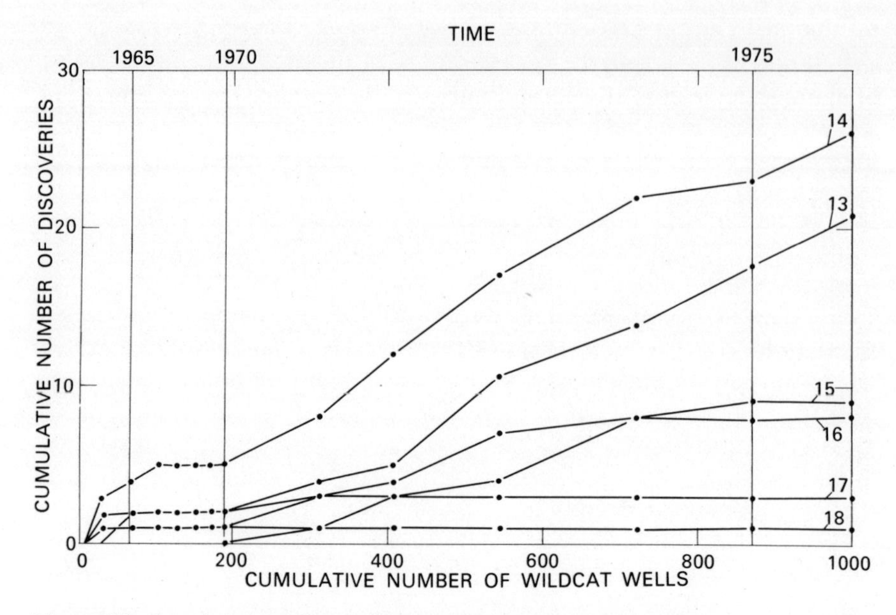

Figure 7.20. Cumulative number of oil and gas fields discovered in size classes 13–18 in water 0–200 m deep in the Pleistocene trend of the study area in the Gulf of Mexico.

Figure 7.21. Cumulative number of oil and gas fields discovered in size classes 9–12 in water 0–200 m deep in the Pleistocene trend of the study area in the Gulf of Mexico.

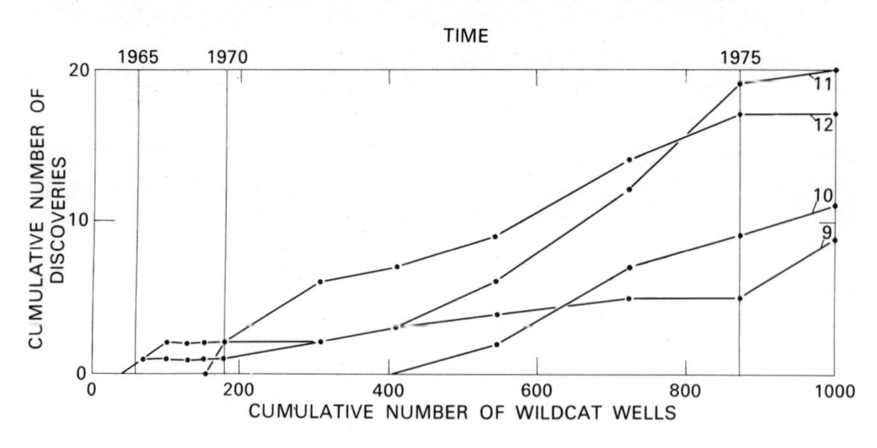

for the total Gulf of Mexico offshore region reveals an improvement in the smoothness of the graphs for size classes 15–18 and an increase in the degree of diminishing returns of wildcat drilling. To the untrained eye, this may not seem like much of a gain for all the effort, but it is such an improvement that the parameter of the Arps and Roberts discovery process model could be estimated directly for the five graphs shown in Fig. 7.17 and several of the graphs in Figs. 7.18–7.21 by using the nonlinear procedure developed by Marquardt (1963). It is obvious, however, that blocking the data by geologic trend is not totally satisfying because the cumulative discovery profiles for field size class 15 still show the effect (upward break) of the change in emphasis of exploration from crude oil toward natural gas (Fig. 7.17). This is a result of the fact that there were enough natural gas prospects remaining to be drilled in the Miocene-Pliocene trend that when drilled during the early 1970s, resulted in the discovery of size class 15 fields. The same upward breaks are also still present in the cumulative discovery rate profiles for most of the smaller field sizes in the Miocene-Pliocene trend (Figs. 7.18 and 7.19).

Fortunately for the progress of the project, my colleague Jack Schuenemeyer was able to join us at this time. He adapted Marquardt's NLIN2 program to our computer, and we solved for the parameters of the Arps and Roberts model (Eq. 6.2) simultaneously for the ultimate number of fields and the efficiency of their discovery. In the Miocene-Pliocene trend, estimates were obtained for field size classes 19 down through 13 (Table 7.2). Below field size class 13, no meaningful fits to the model were obtained for this region. When we tried to estimate the parameters of the discovery process model for the field size class data from the Pleistocene trend, we did not obtain convergence for a single field size class. The data series were short and too erratic (Figs. 7.20 and 7.21). We produced a forecast for this region by using the assumption that the same dis-

Table 7.2. Simultaneous Estimates of the Expected Ultimate Number of Discoveries [$\hat{F}_A(\infty)$] and the Amount of Efficiencies of Discovery [$\hat{C}(A)$] for the Combined Miocene and Pliocene Trend

Size class	Number discovered	$\hat{F}_A(\infty)$	$\hat{C}(A)$
13	50	120.3	2.55
14	42	65.9	4.78
15	43	50.7	4.94
16	25	25.8	3.31
17	17	17.0	5.35
18	5	5.0	[a]
19	2	2.0	[a]
Total	184	286.7	20.93

[a]Data insufficient to estimate $\hat{C}(A)$.

covery efficiency parameter estimates would describe the rates of discovery in the Pleistocene trend as we had obtained for the Miocene-Pliocene trend.

One of the results we obtained from this modeling effort was to relearn the old lesson that it is easy to demand too much from a data series. I realized that the simultaneous estimation of two parameters for each field size class was a heavy demand indeed. This bothered me because the data series of the Gulf of Mexico was the best-prepared data series that we had or ever would have, as far as accuracy was concerned. I looked back at the total-volume-to-total-volume fitting criterion that we had used in the Permian basin study and realized that, however clumsy it was, it produced estimates and it produced them easily.

The disappointment of not obtaining a more complete set of simultaneous solutions was not a permanent defeat. We began to see a light at the end of the tunnel when the graphs in Fig. 7.18 were plotted. I kept going back and looking at these three graphs. I knew that this set of data was trying to tell us something. Why did the discovery rate profile for field size class 13 fall above the graph for field size class 14 and the graph for field size class 12 fall below that for field size class 14? I had seen this happen before and knew that it was connected to the economics of exploration. As I fretted about these three graphs, a key began to turn the tumblers in a lock that led to what was going to be, for me, at least, a startling conclusion.

In only a few days our disappointment at not obtaining a more complete set of direct estimates would shift to enthusiasm. It would take one more piece of graph paper to gain an insight into the nature of the distribution of the parent population of oil and gas field sizes. The next chapter is an exposition of how we stumbled onto a critical ratio. This ratio provided a new twist to the business of estimating the form and the parameters of the parent population of oil and gas field sizes and also the forecasting of future rates of discovery.

8

The Parent Population of Oil and Gas Fields Is Log-Geometric

When we started our work on the Gulf of Mexico, we had hoped that we would be successful in simultaneously estimating the parameters in the Arps and Roberts discovery process model. In Chapter 7, the story was told about how we acquired the data and of our partial success in achieving our goal of simultaneously estimating two parameters in the Arps and Roberts model for each field size class. As mentioned at the end of that chapter, we found out that the data contained other, far more important information. It is the purpose of this chapter to tell the story of the revelation of this information and how it was initially accepted in the fields of discovery process modeling and oil- and gas-resource assessment. The results of the assessment of the undiscovered oil and gas resources in the Miocene-Pliocene and the Pleistocene trends in the Gulf of Mexico are presented at the end of this chapter.

This story really began as we weighed the pros and cons of which fitting criteria should be used to estimate the parameters in the discovery process. We wanted to use a procedure that included the diminishing rate of return information that developed in each field size class as the exploration process evolved over time. However, even with a well-prepared set of wildcat drilling and discovery data of the size we had for the Gulf of Mexico, we were not able to obtain as many simultaneous solutions as we would have liked on which to base our forecast.

Jack Schuenemeyer and I agreed that the Arps and Roberts model was well constructed and contained a reasonable characterization of the discovery process. We also knew that the data with which we had to work did not behave well enough for the simultaneous estimations of two parameters in all the field size classes. We agreed that a major part of the large demand placed on the discovery rate data by the Arps and Roberts model was the result of treating each field size class individually. In the real world, the discovery rate behavior in each field size class is at least vaguely connected to the behavior in every other field size class, and the behavior in adjacent field size classes is closely connected; that is, if you know something about how many discoveries have been made in field size class f_i, then you know something about how many have been made in field size class f_{i+1} and so on for all the size classes. We concluded that this relation could be captured, at least partially, by linking together the estimates of the efficiencies of discovery for each size class and by smoothing them using a running average or a linear fit. A linear linkage already has been suggested from the estimation of the discovery efficiency parameters for the Denver basin (Fig. 6.5).

Linking together the discovery efficiencies caused us to consider how we were going to handle this for the field size classes in the vicinity of the mode of the observed field size distribution. It is at this mode that the simultaneous estimation procedure fails to converge. At this point, the discovery rate profile of the next smallest size class falls below that of the modal class (see Fig. 7.18); that is, something significant occurs at the mode of the observed field size distribution.

The question I asked myself was, is the mode of the observed distribution determined more by the characteristics of the parent population or by the economics of the exploration process? Certainly, the regional geology determined why the few largest fields occur in the area, but what about the great mass of smaller fields that usually occur? Is there a special geology that describes the large occurrences of oil and gas and a common or nearly indistinguishable geology that describes the occurrence of the smaller fields? I had often reached that qualitative conclusion in the past. It was reasonable to believe that in the building of a clastic wedge, lots of local events (for example, local structural events or small reversals in a transgression or a regression that could cause small amounts of petroleum to be trapped) could occur as the wedge passed through the thermal events that cause expulsion and migration of crude oil and natural gas. Such structural and sedimentation events could be so varied and so numerous that it would be unlikely that they could ever be fully described. No formal data analysis had been done on my part; it was just a conclusion that I had reached from my understanding of sedimentation and structural geology combined with an examination of observed field size distributions in a number of regions. I kept coming back to the idea that there had to be an increasing number of smaller and smaller oil and gas fields. I could find no reason for believing that, in general, the size distribution had to have a mode that was a physical consequence of the factors that generated and trapped oil and gas. To be sure, this may have occurred in some regions, but it seems to me to be a most unlikely event.

What else did we know about the mode of the observed field size distribution at that time? It was an accepted fact that a minimum economic field size was calculated and used as an approximate basis for the go or no-go decisions in offshore areas, such as the Gulf of Mexico, where exploration and development costs are commonly high. The bottom-line criterion in this type of decision, which is made by an exploration operator, is based on the anticipated production rate of a field. We could not move effectively toward our goal by trying to work at this level of detail. Our judgment was that we only needed to know the initial production rate of a field, which is the fundamental economic criterion used to make the decision of whether or not to bring a field into production, and that it is proportional to field size.

We also knew from various engineering calculations that the mode of the observed distribution of field sizes was approximately the place in the distribution where fields were beginning to become uneconomic to develop; that is to say, some of the fields that had sizes to the left of the mode were economic to develop, but, on average, fewer and fewer of these fields had production rates that were high enough to offset the cost of exploration and development. Also, after exploration costs have been incurred, it is common to treat them as sunk costs and to go ahead with development if the development costs can be recovered along with some of the sunk costs. These facts imply that the region to the left of the mode diminishes in frequency to the left for purely economic reasons. Arps and Roberts (1958) specifically stated that they believed this to be the reason for the diminishing left-hand tail of the size distribution for the Denver basin.

Although the idea that the observed field size distribution is subject to economic truncation in the vicinity of its mode is not a new one, we did face an institutional problem as we started to work with it. The problem revolved around the tradition that the lognormal distribution was the accepted distribution for the parent population of oil and gas field sizes. Under static conditions, it usually fit the observed data very well. Because this distribution was difficult to work with analytically, it required the expenditure of much intellectual effort to achieve a structure that would be useful for summarizing and forecasting.

We knew that approaching the mode of the observed distribution of oil and gas fields from the angle that it was an artifact of the economics of the exploration process would require some courage. It was a little more troublesome than it would otherwise have been because we respected the work of Gordon Kaufman, even though we were about to strike at the heart of his position. Kaufman's work on characterizing the exploration play was not at issue (see Chapter 4); rather, it was his choice of the lognormal distribution. He had been the main protagonist in the use of the lognormal distribution as a model for the parent population of oil and gas field sizes. Putting our respect for the man's enthusiasm and his pleasant manner aside, we began our attack.

Our idea was to rummage through all our data sets to determine if any systematic pattern could be found that would provide a basis for getting around the problems introduced into the discovery time series by the economic truncation of the data. Some of our qualitative data suggested that the truncation phenomenon was nontrivial. When we were building the data files for the Gulf of Mexico offshore region, Walt Bawiec and I had found evidence of substantial releasing of previously drilled leases; by the late 1970s, some 20 leases had been dropped twice and were under lease again for the third time, and more than 70 leases had been released a second time. Inspection of the reserve estimates for these blocks showed that only small amounts of oil and, more commonly, natural gas occurred in these blocks. I could only conclude that this amount of releasing indicated the presence of a relatively large number of oil and gas fields whose field sizes were in the vicinity of the mode of the observed field size distribution. These fields had not been economic to develop in the past, but, as crude oil and natural gas prices began to rise in the 1970s, their potential was reviewed.

The releasing of previously leased and dropped blocks was solid evidence, but it could not by itself be cast in a systematic form to be used as a basis for modeling our way through the barrier erected by the economic truncation of the observed field size data. However, it was from such data that we began to conclude that, maybe, the majority of the oil and gas fields remaining to be discovered in a partially explored area had sizes in the vicinity of or below the mode of the observed size distribution for the fields already discovered.

Our search for evidence of economic truncation in our discovery rate data led us to reexamine the Permian basin data sets where we found another useful piece of information: The modes of the observed distribution migrated outward as the depth interval increased (as cost increased). We also noticed that the mode of the observed distribution in the Gulf of Mexico offshore was even larger than the mode in the 10,000- to 15,000-foot interval in the Permian basin. Too few discoveries had been made in the 15,000- to 20,000-foot interval in that basin to construct a meaningful observed field size distribution.

Next, we fit the Arps and Roberts model to each of the discovery rate profiles within each of the Permian basin depth intervals by using the simultaneous estimation procedure that had been used on the Gulf of Mexico data. The data fit the model closely. We plotted the estimated ultimate number of fields in each size class on top of the observed distribution. Jack Schuenemeyer then made a critical graph, which is shown in Fig. 8.1. The data in this graph, along with subsequent analyses, yielded a solution to the economic truncation problem. Jack, who had drawn this graph in his office at the University of Delaware, said nothing to me about it before dropping it in the mail. Standing in front of my drafting table on a cold and snowy February morning in 1980, I opened

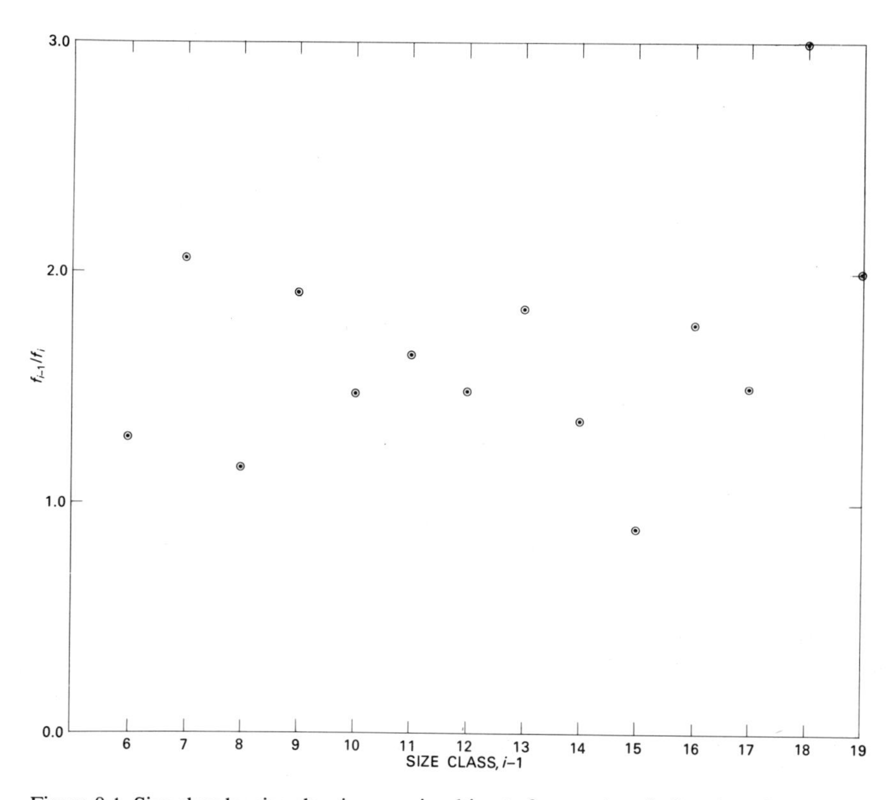

Figure 8.1. Size-class-by-size-class increase in ultimate frequencies of oil and gas fields in the 0- to 5000-foot depth interval of the Permian basin. The average rate of increase is F_{i-1}/F_i = 1.67 (Drew et al., 1982).

the brown envelope and took out the graph along with some other tabular material. I went immediately to the phone to call Jack, exclaiming, "Do you know what this f_{i-1} versus f_i graph could mean to us?" His response was that he did know what it could mean. Later, when I asked why he made this graph, he said that he had just plotted anything and everything that made any kind of sense.

The information contained in this graph meant that, after estimating the ultimate number of fields expected to occur in each field size class ($F_A(\infty)$ in Eq. 6.2) above the mode in the 0- to 5000-foot depth interval in the Permian basin, the expected frequency of occurrence of fields in each size class was a constant multiple of the expected number in the next larger size class. This conclusion would be a powerful cure for the ills introduced into the observed discovery rate data for the field size classes in the vicinity near and below the mode of the observed field size distribution by economic truncation if we could show that the f_{i-1} to f_i size class ratio was always a constant above the mode of the observed field sizes and that the mode migrated systematically with, say, the cost of drilling a wildcat or development well.

During the following several weeks, we determined that even though f_{i-1}/f_i varied widely in those field size classes containing only a few discoveries, there was no evidence to overturn the original conclusion deduced from the data displayed in Fig. 8.1. We demonstrated that the mode of the observed distribution of oil and gas fields migrated upward as a function of the cost of exploration by comparing the position of the modal field size class with the associated cost of drilling a wildcat well. The Permian basin data set yielded three data points to this relation, one from each of the upper three depth intervals, and the Gulf of Mexico provided a fourth. The modes of these four respective observed field size distributions were found to occur in size classes that were at least ordinally related to the cost of drilling. A fifth data point was generated from the observed field size distribution for the Denver basin data set. When this distribution was drawn to the same scale, its mode was found to be in a larger size class than the mode of the 0- to 5000-foot depth interval in the Permian basin. Could the cost of drilling a wildcat well into the 0- to 5000-foot interval in the Permian basin be lower than drilling a wildcat well to the average depth of exploration in the Denver basin? I did not think it likely because the average well in the Denver basin was drilled in 3 or 4 days by using a couple of shale bits to reach the target depths that were usually between 5000 and 7000 feet. The Permian basin had a lot of sandstones, limestones, and dolomites, which I suspected would make the cost of drilling to an average depth of around 3000 feet in the 0- to 5000-foot interval in the Permian basin to be about the same as in the Denver basin. Inspection of the drilling cost data in the 1972 "Joint Association Survey," the most recent issue available to us, revealed that the approximate cost of drilling an average wildcat well in 1972 in the Denver basin was $43,000, which was nearly twice the cost of drilling the same type of well in the 0- to 5000-foot depth interval in the Permian basin (Table 8.1). So, we gained another data point that contributed further ordinal consistency to the relation between the position of the mode of the field size class and the cost of drilling.

By using the above analysis, we concluded that the mode of the observed field size distribution was linked directly to the cost of drilling a wildcat well. Combining this idea with the relation shown in Fig. 8.1, we asserted that we had solved the fundamental problem in oil- and gas-resource assessment, which was to estimate the parameters of the parent field size distribution. I recognized that our data analysis had driven us to a solution along a path that is closely related to the solution of the well-known "identification problem" in econometrics. This solution consisted of recognizing that in the cross section of exploration plays, basins, and countries, the mode of the observed distribution migrated inward as the cost of drilling fell. As the mode moved, it traced out a path of economic truncation, which, in turn, allowed us to project the constant $r=f_{i-1}/f_i$ through the truncation point, which yields an estimated number of

Table 8.1. Approximate Cost of Wildcat Drilling in 1972

Area	Cost[a] (in U.S. dollars)	Modal class	Mean size in modal class (in 10^3 BOE[b])
Permian basin:			
0- to 5000-foot interval	26,000	6	136
5000- to 10,000-foot interval	74,000	7	272
10,000- to 15,000-foot interval	275,000	11	4,170
Denver basin average cost	43,000	8	537
Gulf of Mexico average cost	564,000	13	18,050
World offshore average cost	2,000,000	16	122,000

[a]Dry-hole cost used to approximate the cost of a wildcat well.

[b]BOE = barrels of oil equivalent.

Source: American Petroleum Institute, Independent Petroleum Association of America, Mid-Continent Oil and Gas Association (1972). Modified from Schuenemeyer and Drew (1983).

fields in each field size class below the truncation point in the observed field size distribution. By putting this two-stage estimation scheme in place, we were asserting that the mode of the observed distribution is an artifact of the cost of exploration and, more importantly, that the underlying parent population of oil and gas fields in any region is distributed as a log-geometric distribution. This distribution belongs to the family of distributions that is commonly known as "J-shaped," or Pareto.

A graphic exposition of the elements in this two-stage estimation scheme is shown in Figs. 8.2–8.6. First, it is useful to show the type of forecast that is produced when the truncation effect is *disregarded* (top profile, Fig. 8.2). The point of truncation is labeled S_0. To the right of this point, the fields are large enough, on average, to be economic, and, to the left, they are not. Although the undiscovered portion of the distribution to the right of the truncation point can be estimated correctly by using a discovery process model, the portion to the left will be underestimated, and the mode has been introduced as an artifact of the economics of exploration. This does not mean that such a forecast is totally worthless. If static economic conditions were to prevail, then such a forecast would be of considerable use; for example, when we developed our discovery process model based on the concept of the area of influence of a drill hole, we made a forecast for the Denver basin (Fig. 8.3). By using a back-forecast as a test criterion, we found it to be useful for the time period of the forecast. During that time period, conditions of nearly static costs and prices prevailed. The most striking aspect of this forecast (Fig. 8.3) is that the discovery process model forecasted not only the actual number of discoveries closely in those size classes above the point of economic truncation, but also that below the mode and on down through very small field size classes.

This analysis cleared up the mystery about why the lognormal distribution model was so widely accepted as a model for the parent population of oil and

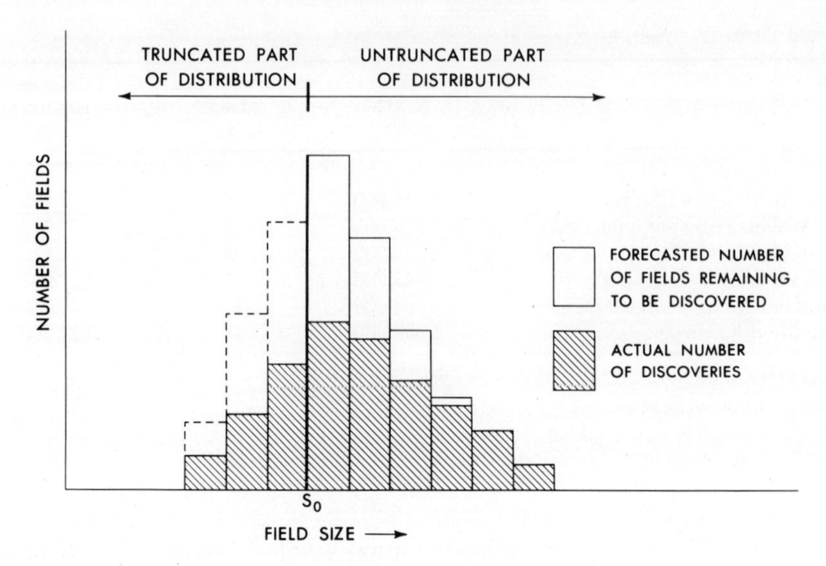

Figure 8.2. Actual discoveries and a forecast of the ultimate number of fields occurring in each size class for which no corrections for economic truncation have been made. S_0 is the point of economic truncation.

gas fields. Under static economic conditions, it is a good respresentation because, as the data in Fig. 8.3 show, it is a product of this very condition.

For my own account, I can say that I was aware that Arps and Roberts mentioned the idea of economic truncation. Later, at least a few of us knew that some sort of truncation effect was occurring. None of us, however, had systematically put the ideas together until the demand for a forecast of future rates of discoveries in the Gulf of Mexico offshore forced the issue and a solution was unavoidable.

From the ideas presented above, we need to stop only briefly at Fig. 8.4 before going on to a discussion of the data presented in Figs. 8.5 and 8.6. The profiles shown in Fig. 8.4 schematically illustrated the contrast between the bias introduced by ignoring economic truncation and the companion forecast when this effect is considered. This latter forecast is equivalent to assuming that the parent population of oil and gas fields is log-geometric. The point of this figure is that any forecast made in which the effect of truncation is ignored will become progressively less and less accurate in smaller and smaller field size classes. This will be an important concept when we discuss the undiscovered oil and gas resources of the Gulf of Mexico offshore later in this chapter.

Now we turn to a discussion of the estimation of the number of fields remaining to be discovered in the truncated part of the observed field size distribution. The first task is to solve $\hat{r} = \hat{f}_{i-1}/\hat{f}_i$ for the untruncated portion of the observed field size distribution as displayed schematically in Fig. 8.5. The pat-

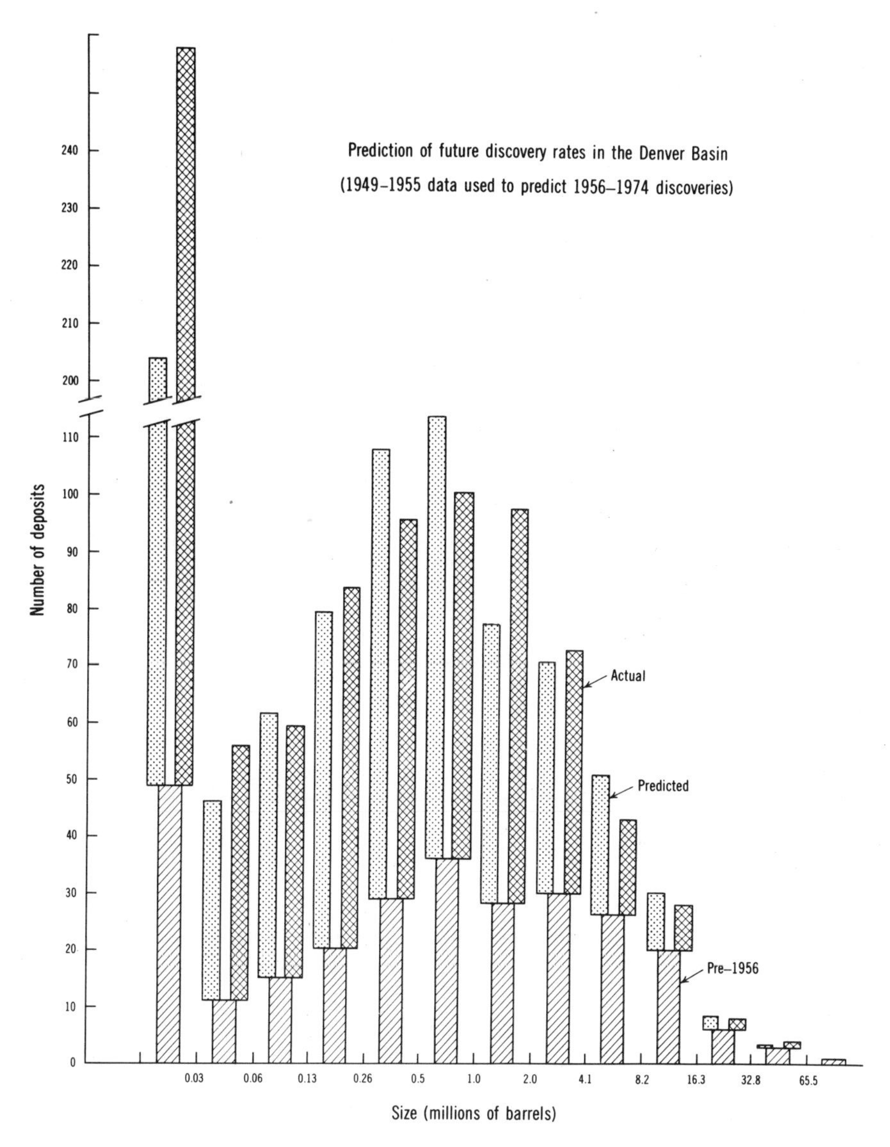

Figure 8.3. Data for exploration in 1949–1955 used to predict numbers of wells discovered in 1956–1974 for the Denver basin.

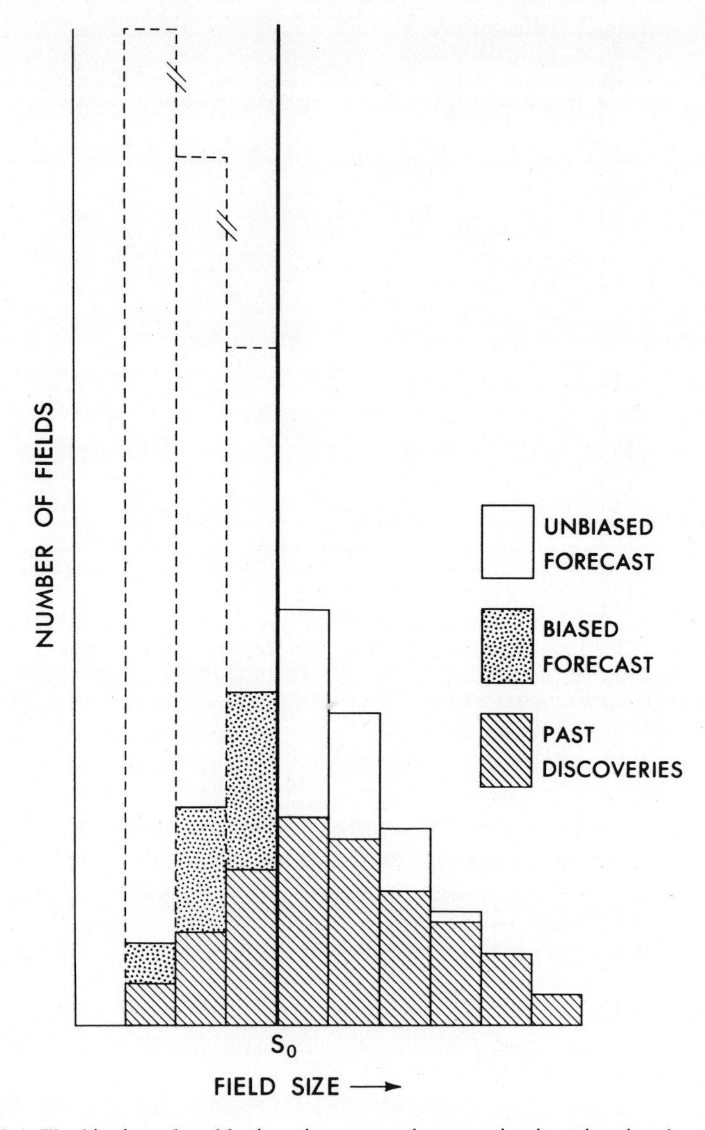

Figure 8.4. The bias introduced by ignoring economic truncation in estimating the ultimate number of fields occurring in each size class and the postulated form of an unbiased forecast. S_0 is the point of economic truncation.

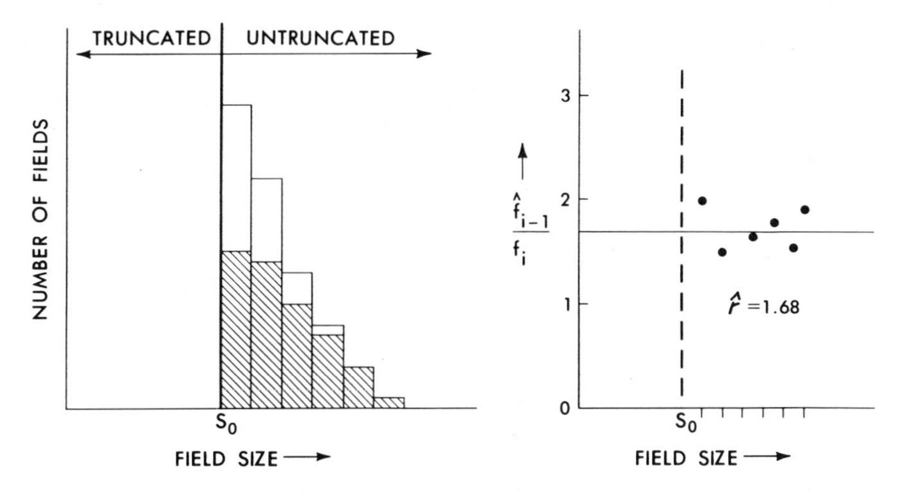

Figure 8.5. The procedure used to estimate the r factor from the untruncated portion of the observed field size distribution. Left, observed (pattern) and predicted (no pattern) numbers of discoveries in field size classes larger than S_0. Right, graph of the ratios of numbers of discoveries in adjacent field size classes larger than S_0, where S_0 is the point of econimic truncation.

terned portion of the distribution displayed in the left panel of Fig. 8.5 represents the observed field size distribution. The top profile in this panel represents the ultimate number of fields expected to occur in each field size class. The data from this upper profile are used to solve $\hat{r} = \hat{f}_{i-1}/\hat{f}_i$ in the panel on the right in Fig. 8.5. This estimate of $\hat{r}$ is then used to forecast the expected number of fields occurring in each field size class over the size range in which truncation has occurred. This estimation process is illustrated in Fig. 8.6. A factor of $\hat{r}$ = 1.68 has been used to estimate the number of fields expected to occur in each size class in the zone of economic truncation. The patterned frequency distribution shown in this figure schematically represents the forecast that would be made for the ultimate number of fields expected to occur if economic truncation is ignored. The shaded portion of the distribution schematically represents a forecast that accounts for the effect of economic truncation.

This procedure was used to estimate the size of the undiscovered resource base for the Miocene-Pliocene and the Pleistocene trends in the Gulf of Mexico offshore (Fig. 7.2). The summary data for the ultimate expected number of discoveries in the Miocene-Pliocene trend are shown in Table 8.2. From this table, it can be seen immediately that the number of fields in field size classes 12 down through size class 9 estimated to be undiscovered in this combined trend is relatively large. The observed frequency distribution of discoveries used to identify the truncation point is shown in Fig. 8.7. When we made this forecast, we knew that we would have to defend ourselves because we were saying that there were nearly 2000 oil and gas fields in the size range from 0.76 million

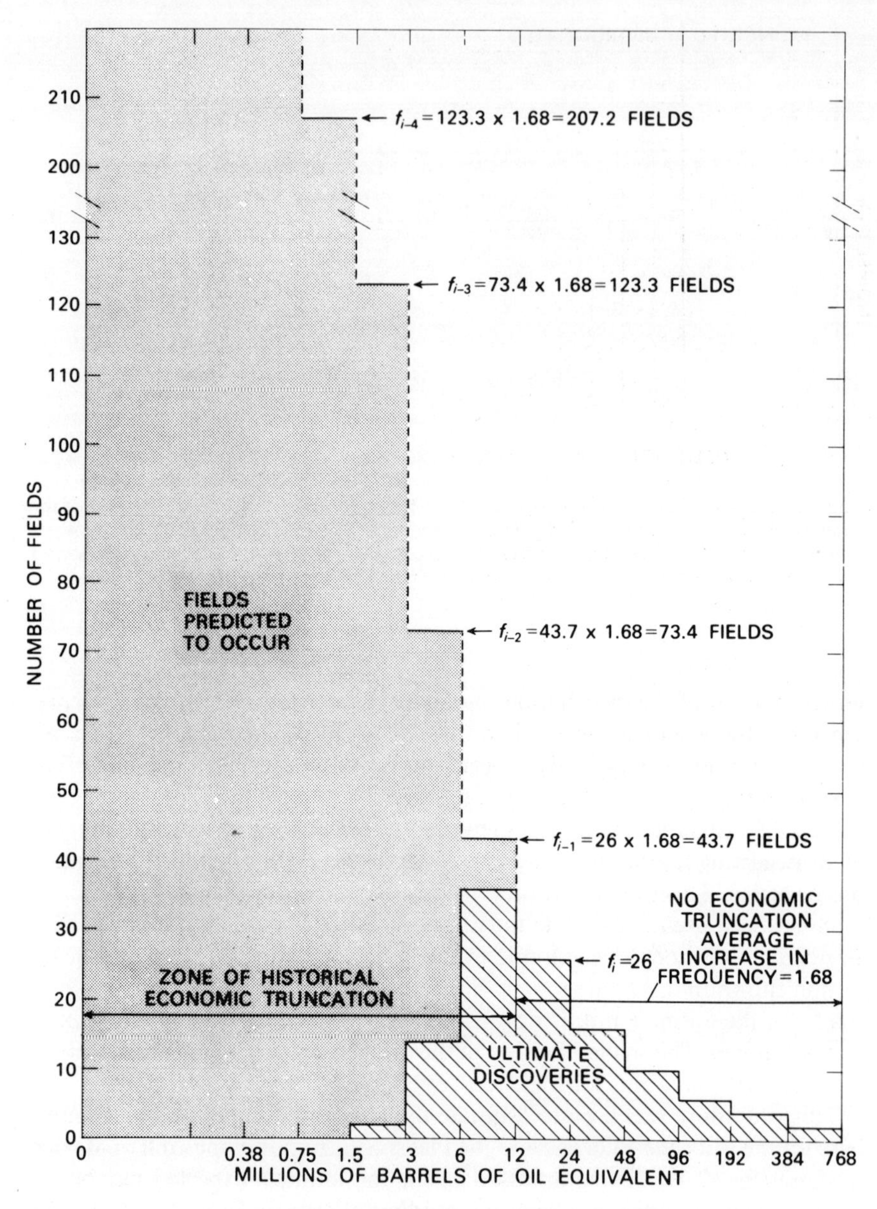

Figure 8.6. The method used to estimate the part of the field size distribution below the historical level of economic truncation. Cross hatched distribution, as shown by pattern, results when economic truncation effect is ignored.

Table 8.2. Summary Data and Estimates of the Ultimate Number of Oil and Gas Fields by Class Size and Contained Resources for the Combined Miocene and Pliocene Trend of the Study Area

Size class[a]	Number of fields found by 3100 wildcat wells	$\hat{F}_A(\infty)$[b]	Unsmoothed $\hat{C}(A)$	Smoothed $\hat{C}(A)$	Expected total ultimate resource (BOE[c] $\times 10^4$)	Expected remaining resource[d] (BOE $\times 10^4$)	Percent of total resource remaining
9	24	889.4	—	—	1005.0	977.9	12.7
10	27	539.1	—	—	1223.8	1162.5	15.1
11	27	326.7	—	—	1509.4	1384.6	18.0
12	35	198.0	—	—	1716.7	1413.2	18.3
13	50	120.3	2.55	2.65	2223.1	1299.1	16.9
14	42	65.9	4.78	3.40	2282.1	827.7	10.7
15	43	50.7	4.94	4.15	3551.0	539.3	7.0
16	25	25.8	3.31	4.89	3302.7	102.4	1.3
17	17	17.0	5.35	5.65	4583.7	0	0
18	5	5.0	[e]	[e]	2873.2	0	0
19	2	2.0	[e]	[e]	2273.7	0	0
Total	297	2,239.9	—	—	26,544.4	7,706.7	100.0

[a]See Chapter 6, Table 6.2, for definition of field size classes.

[b]The ultimate numbers of fields estimated to exist in size classes 13–19 were estimated directly from the discovery data series. The ultimate numbers of fields is size classes 9–12 were estimated by applying the average $f_{i-1}/f_i = 1.65$ ratio determined from size classes 13–17.

[c]BOE = barrels of oil equivalent.

[d]Amounts are for the period starting 1/1/77.

[e]Insufficient data to estimate $\hat{C}(A)$.

Note: Dash leaders indicate that the nonlinear least-squares procedure could not be used; BOE, barrels of oil equivalent; $\hat{F}_A(\infty)$, estimated ultimate number of fields in a size class; $\hat{C}(A)$, estimated efficiency of exploration for size class.

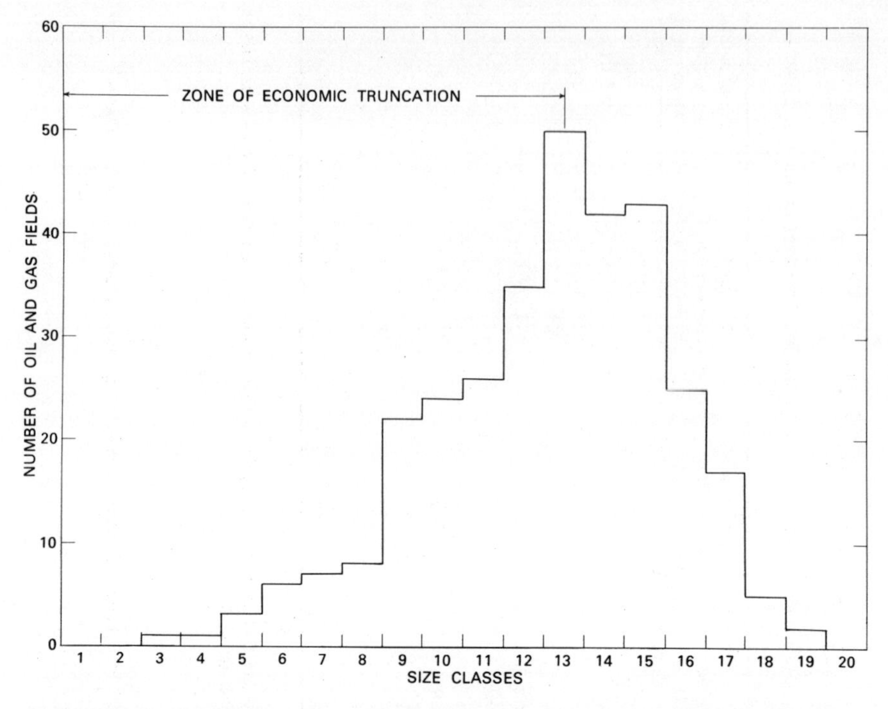

Figure 8.7. Size distribution of oil and gas fields discovered in the combined Miocene and Pliocene trend of the study area in combined state and federal waters 0–200 m deep as of December 31, 1976.

to 12.14 million barrels of oil equivalent remaining to be discovered in the Miocene-Pliocene trend.

This conclusion easily could have been too revolutionary to gain acceptance. Before these results were presented, nobody talked about the idea that lots of little oil and gas fields remained to be found in the partially explored basins in the United States or anywhere else in the world. That this could be true and that you could estimate how many there might be and what proportion of the total undiscovered oil and gas they contained was, to say the least, an experimental idea. To satisfy myself, I needed to talk to somebody who was watching exploration operations in the Gulf of Mexico offshore and to ask if we had reached a reasonable conclusion. My first call was to Rod Pearsy, the manager in the then Conservation Division of the USGS in Metairie whom Walt Bawiec and I had met during our trip to that office to collect data for this study.

I asked him if there could be as many as 2000 oil and gas fields remaining to be discovered in the Gulf of Mexico which contained between 0.76 million and 12.14 million barrels each and if about one-half of these fields could range in size from 0.76 million to 1.5 million barrels of oil equivalent. His answer was that he could imagine that many occurring in the shale platforms alone.

He believed that many small stratigraphic traps existed that might contain such small amounts of oil and gas. I took these remarks as support for our conclusion from a geologist who was aware of the status of the petroleum geology in the Gulf of Mexico—not in the sense that we were right but in the sense that we were not totally off base.

Our conclusion did not go unchallenged for long, however. About the time that we were putting the finishing touches on our soon-to-be-published first results of the Gulf of Mexico offshore forecast (Drew et al., 1982). I was confronted by a very unhappy visitor. His name was Richard Nehring. He had come to see Dick Meyer, my supervisor, to complain about the new idea. It was Meyer's opinion that any new idea should be bashed about as soon as possible to assess its value. Because Nehring was one of the better known characters in the petroleum-resource assessment business, Meyer thought that it would be useful to have someone of Nehring's stature give his views on our work; it would also help firm up his own opinion on the value of what we had done. Nehring had made extensive tabulations on the characteristics of the largest oil and gas fields in the world. From the data in these tabulations, he had made the observation that most of the oil and gas in the world was contained in the few largest fields and that most of, if not all, these fields had been discovered already and were now being depleted rapidly. From this conclusion, he went on to produce a pessimistic forecast for the volume of oil and gas that would be discovered in the future in the United States and in other regions in the world.

When I gave him a briefing on our technique and a preview of the results from the Miocene-Plocene trend (Table 8.3), he got mad. There is no other way to say it; he just got mad. He erupted at the idea that the majority of the oil and gas left to be discovered in the combined trend could be in fields smaller than the mode. He also really did not like the forecast that 8 billion barrels of oil equivalent remained to be discovered in the combined trend and that it was mostly left to be discovered in 2000 small fields.

He demanded that Dick Meyer write a cease-and-desist order against our work! Nehring left in a disturbed state, only to return about a month later. This time, Meyer called me into his office, Nehring made a charge that what we had done was nonsense, absolute nonsense! At the blackboard, I drew graphs and pushed the chalk with some vigor. I made reference to the migration of the mode of the distributions (shown in Fig. 8.8) and the companion cost data (shown in Table 8.1). I trotted out the newly prepared figure for the observed distribution for the world offshore with its mode in field size class 15 (Fig. 8.9). This mode was shifted outward to the right of the mode for the Gulf of Mexico. The cost of drilling was even higher here than in the Gulf of Mexico offshore. I made the point that, as sure as God made little green apples, He did not make the modes of these observed distributions back in the Silurian, Jurassic, and Miocene times to vary proportionally with the cost of drilling in the year 1972.

As I made my points at the blackboard, I turned from time to time to watch

Table 8.3. Forecasted Future Number of Discoveries per Increment of 200 Wildcat Wells Drilled in the Combined Miocene and Pliocene Trend of the Study Area

Cumulative number of wildcat wells	Number of fields to be discovered in each class[a]								Incremental discoveries in all size classes (millions of BOE[b])
	9	10	11	12	13	14	15	16	
3300	7.78	5.77	4.36	3.26	2.39	1.51	0.88	0.16	251.1
3500	7.70	5.70	4.28	3.19	2.31	1.42	0.78	0.13	233.0
3700	7.62	5.62	4.21	3.11	2.23	1.33	0.69	0.10	217.2
3900	7.54	5.55	4.13	3.03	2.16	1.24	0.61	0.08	203.2
4100	7.47	5.48	4.06	2.96	2.08	1.16	0.54	0.06	190.7
4300	7.39	5.40	3.99	2.89	2.01	1.09	0.48	0.05	179.5
4500	7.31	5.33	3.92	2.82	1.95	1.02	0.42	0.04	169.4
4700	7.24	5.26	3.85	2.75	1.88	0.96	0.38	0.03	160.2
4900	7.16	5.19	3.78	2.69	1.81	0.90	0.33	0.03	151.9
5100	7.09	5.13	3.72	2.62	1.75	0.84	0.29	0.02	144.3
5300	7.02	5.06	3.65	2.56	1.69	0.79	0.26	0.02	137.4
5500	6.95	4.99	3.59	2.50	1.64	0.74	0.23	0.01	131.0
5700	6.87	4.93	3.53	2.44	1.58	0.69	0.20	0.01	125.1
5900	6.80	4.86	3.46	2.38	1.53	0.65	0.18	0.01	119.7
6100	6.73	4.80	3.40	2.32	1.47	0.61	0.16	0.01	114.6
6300	6.67	4.73	3.34	2.27	1.42	0.57	0.14	0.01	109.9
6500	6.60	4.67	3.28	2.21	1.38	0.53	0.13	0.00	105.6
6700	6.53	4.61	3.23	2.16	1.33	0.50	0.11	0.00	101.5
6900	6.46	4.55	3.17	2.11	1.28	0.47	0.10	0.00	97.6
7100	6.40	4.49	3.11	2.06	1.24	0.44	0.09	0.00	94.0

[a]See Chapter 6, Table 6.2, for definition of field size classes.
[b]BOE = barrels of oil equivalent.

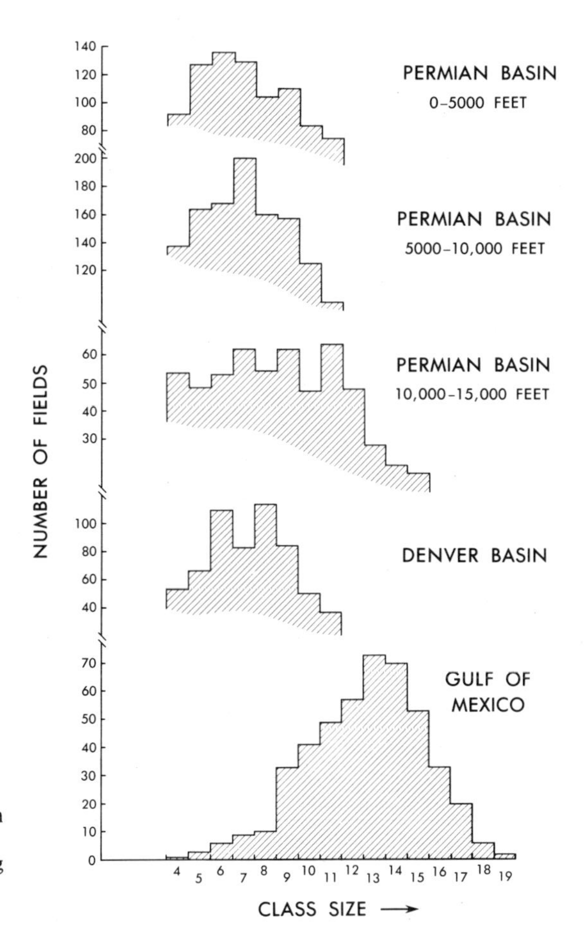

Figure 8.8. The modal shift in the field size distribution as a function of the cost of drilling (Schuenemeyer and Drew, 1983).

Nehring. He was either looking out the window or rustling papers on the table in front of him and definitely was not paying attention. Meyer silently watched the interchange. Afterward, he invited Nehring to join me and Walt Bawiec in my office to get down to the technical meat of the issue. At my drafting table, I laid out a lease-block map and a computer listing of the lease-block assignments that had been made over the last 20 years in the Gulf of Mexico. I then tried to explain the phenomenon of releasing acreage; the level of releasing indicated that small fields had been discovered and dropped that contained reserves near the size of the mode of the observed distribution. As I was building my case, Nehring wrenched from my hand the pencil that I had been using for a pointer; with the pencil in his hand, he gained control of the discussion. From that point on, he argued strenuously against my every statement. Walt said to me later that he thought Nehring was pretty emotional.

My relation with Nehring did not end at that point. He later would attack

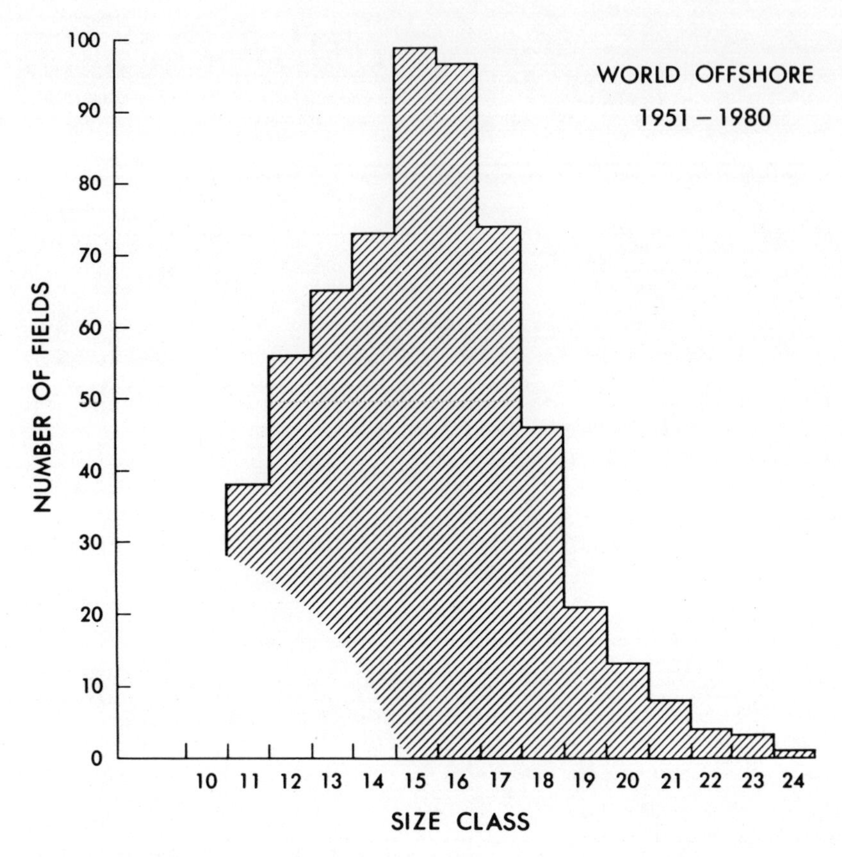

Figure 8.9. The observed size of oil and gas fields in barrels of oil equivalent in the world offshore, 1951–1980.

our estimation of the r factor. He said that his analysis showed conclusively that the value of $\hat{r}$ could not be equal to 1.68 or anything near that large. His calculations showed that the factor, at the outside, could not be larger than 1.2. In looking over his tables, it was obvious that he had missed the step in the estimation procedure where the $F_A(\infty)$ values are estimated size class by size class and then are used to estimate $\hat{r}$. Nehring had calculated his version of the factor from the observed data; that is, $F_A(w)$ values. This oversight would, of course, produce a much smaller estimate of $\hat{r}$. Could it be that Nehring really believed that all the fields larger than the mode had been discovered already or was it a clumsy mistake? I never did find out. He would never answer my direct question on this point.

In early spring 1982, Nehring convened a meeting on oil- and gas-resource assessment under the auspices of his employer, the Rand Corporation. At this meeting, Nehring and I stood head to head and toe to toe and slugged it out in front of a hundred or so of our kind. It is characteristic of the resource-assessment profession that, at one time or another, nearly everybody meets nearly everybody else who works in the field. So it should not be viewed as odd that Nehring convened a meeting and invited me to say why I thought he was wrong. After the meeting ended, I waited eagerly for the publication of the oral transcript. It would have made a dandy reference to cite from, but, unfortunately, it was never published. It is my opinion that, from 1982 onward, Nehring still did not care much for our work but could only point to a few petroliferious regions where it might not apply very well. My response has been a feeling of gratitude to him for pointing out such areas.

In the remainder of this chapter, a technical discussion is presented that uses the actual forecast made of the number of undiscovered fields in the Miocene-Pliocene and the Pliestocene trends and the forecast of their future rates to expose the mechanics involved in making these forecasts for the truncated and untruncated parts of an observed field size distribution. As a point of departure, the data in Table 8.2 show that after January 1, 1977, no fields were expected to be discovered above size class 17; in other words, we did not expect to see any more fields discovered that contained more than 194.3 million producible barrels of oil equivalent (or 1.17 trillion cubic feet of gas equivalent) in the trend. The largest volume of undiscovered oil and gas occurring within a single field size class (18.3 percent of the total) was predicted to exist in field size class 12; that is, within fields that ranged in size from 6.07 million to 12.14 million barrels of oil equivalent or, equivalently, from 36.4 billion to 72.8 billion cubic feet of producible gas. According to the prediction, 1.4 billion barrels of oil equivalent remained to be discovered in the 163 fields in this size class.

In addition, Table 8.2 shows that in the Miocene-Pliocene trend, 64.1 percent of the oil and gas remaining to be discovered in fields larger than 0.76 million barrels of oil equivalent, as of January 1, 1977, was expected to occur in the four smallest size classes. Fields within this size range were below the mode of the observed field size distribution and, therefore, were within the size range in which economic truncation is effective. In spring 1980, when these results were being produced, we believed that the rapidly escalating prices of natural gas and crude oil were going to make many of the fields in this size range attractive economic targets. The importance of this conclusion is that the majority of the oil and gas remaining to be discovered in the combined trend down to a reasonable cutoff field size was not only in small fields, but also in fields that were in the historic zone of economic truncation.

Given the estimates of the ultimate number of oil and gas fields expected to occur in each field size class, along with such necessary data as the average areal extent of the fields, we were able to calculate the future rates of discovery by using the Arps and Roberts discovery process model; this forecast is shown in Table 8.3. The number of discoveries that were expected to be made in each field size class by the first increment of 200 wildcat wells drilled starting on

January 1, 1977, is given in the first row of the table. These oil and gas fields (26.1 in all) were predicted to contain 25.1 million barrels of oil equivalent.

The number of fields that were forecast to be discovered by the first 200 wildcat wells drilled after 1976 ranged from a maximum of 7.78 discoveries expected in field size class 9 down to only 0.16 expected discovery in field size class 16. According to our prediction, the quantity of oil equivalent expected to be discovered by the completion of each drilling increment would decline slowly. Our forecast indicated that almost as many wildcat wells (79.2 percent) would have to be drilled in the future (2400 wells starting on January 1, 1977) as had been drilled in the past (3100 wells through the end of 1976) before the expected level of discovery per drilling increment declines by approximately one-half (125.1 million barrels of oil equivalent per increment, or 625,500 barrels of oil equivalent per wildcat well). If the industry continues to drill several hundred wildcat wells per year in the combined trend, we will be into the early 1990s before the discovery rate falls to 625,500 barrels of oil equivalent per wildcat well. At the time this forecast was made, it was reasonable to conclude that if natural gas and crude oil prices held or continued to increase, this level of discovery per well would be well within reach economically.

As was discussed in Chapter 7, the discovery rate data series that we used for the Pleistocene trend (Fig. 7.2) covered too short an interval of drilling (Figs. 7.20 and 7.21) to allow direct estimation of the ultimate number of fields occurring in each size class or the efficiency of their discovery. Given this situation, we had two choices. Our first option was to make an aggregate discovery rate extrapolation and then to associate a field size distribution at various points along the extrapolation. Our second option assumed that the shape of the discovery rate profiles would mimic those for the the Miocene-Pliocene trend; that is, we could assume that the discovery efficiencies for each field size class in the Pleistocene trend were similar to those estimated for the Miocene-Pliocene trend. We chose the latter option.

We also assumed that the parent population of fields in the Pleistocene trend for the field size classes smaller than 12 had the same general form as the parent population of field size in the Miocene-Pliocene trend, so that $\hat{r} = f_{i-1}/f_i =$ 1.65 for field size class 12 downward to field size class 9.

Having made these two assumptions, the steps in the estimation procedure were set in place. The ultimate number of fields expected to occur in field size classes 13–17 were estimated by solving the Arps and Roberts equation, given the observed data and the assumed discovery efficiencies (Drew et al., 1982, p. 23); for example, in field size class 13, we have

1. $C(A) = 2.65$ (from the combined Miocene and Pliocene trend)
2. Given the additional data:

$A_{13} = 6.68\ \text{km}^2$

$w = 1000$ wildcat wells by the end of 1976

$F_{13}(w) = 21$ fields discovered by the end of 1976, and

$B = 33{,}612\ \text{km}^2 =$ size of the Pleistocene trend

3. Solve for $F_A(\infty)$:

$$F_A(w) = \hat{F}_A(\infty)(1 - \exp(-C(A)Aw/b)).$$
$$21 = \hat{F}_{13}(\infty)(1 - \exp((-2.65 \times 6.68 \times 1000)/33{,}612)):$$
$$\hat{F}_{13}(\infty) = 51.3 \text{ fields expected in field size class 13}$$

Given estimates of the $F_A(\infty)$'s made in this manner for field size classes 13–17, we then used the r factor estimated for the Miocene-Pliocene trend to make the corresponding estimates for field size classes 12 down through 9.

The results obtained for the Pleistocene trend by using this estimation procedure are displayed in column 3 of Table 8.4. A total of 901.6 oil and gas fields were expected to occur in the Pleistocene trend, each containing more than 0.76 million barrels of producible oil equivalent. Of this total, 125 had been discovered by the end of 1976. The expected total productivity of the 776.6 fields remaining to be discovered in the Pleistocene trend was estimated to be 2.95 billion barrels of oil equivalent. Thus, about 40 percent of the ultimate productivity of the Pleistocene trend remains to be discovered after January 1, 1977, whereas only 29 percent of the producible oil and gas remaining to be discovered in the Miocene-Pliocene trend remains to be discovered after the same date. Comparison of the results in Table 8.4 with those in Table 8.2 shows, as should be expected from the use of the log-geometric distribution, that the proportions occurring by field size class are nearly identical, although the total volume of oil and gas expected to be discovered in the Pleistocene trend is much less than in the Miocene-Pliocene trend.

Table 8.4. Summary Data and Estimates of the Ultimate Number of Oil and Gas Fields by Class Size and Contained Resources for the Pleistocene Trend of the Study Area[a]

Class size[b]	Number of fields found by 1000 wildcat wells	$\hat{F}_A(\infty)$[c]	Smoothed $\hat{C}(A)$	Expected total ultimate resource (BOE × 10^4)	Expected remaining resource[d] (BOE × 10^4)	Percent of total resource remaining
9	9	358.7	—	391.0	381.2	12.9
10	11	217.4	—	450.0	427.3	14.5
11	20	131.8	—	581.2	493.0	16.7
12	17	79.9	—	670.4	527.7	17.9
13	21	51.3	2.65	871.6	514.8	17.5
14	26	39.9	3.40	1392.1	485.0	16.5
15	9	10.4	4.15	692.7	93.3	3.2
16	8	8.2	4.89	1038.9	25.3	0.9
17	3	3.0	5.65	623.4	.0	0.0
18	1	1.0	—	502.8	.0	0.0
Total	125	901.6	—	7214.1	2947.6	100.0

[a] $\hat{F}_A(\infty)$, estimated ultimate number of fields in a size class; $\hat{C}(A)$, estimated efficiency of exploration for size class; BOE, barrels of oil equivalent.

[b] See Chapter 6, Table 6.2, for definition of field size classes.

[c] See text for explanation of estimation procedure used.

[d] Amounts are for the period starting January 1, 1977.

Table 8.5. Forecasted Future Number of Discoveries per Increment of 200 Wildcat Wells Drilled in the Pleistocene Trend of the Study Area

Cumulative number of wildcat wells	Number of fields to be discovered in each class[a]								Incremental discoveries in all size classes 9–16 (millions of BOE[b])
	9	10	11	12	13	14	15	16	
1200	5.84	5.18	4.50	3.73	3.03	2.64	0.46	0.11	256.8
1400	5.73	5.04	4.31	3.49	2.73	2.14	0.31	0.05	213.1
1600	5.63	4.90	4.12	3.26	2.45	1.73	0.21	0.02	180.9
1800	5.53	4.77	3.95	3.04	2.21	1.40	0.14	0.01	156.0
2000	4.53	4.64	3.78	2.84	1.99	1.14	0.09	0.01	136.4
2200	5.34	4.51	3.62	2.66	1.79	0.92	0.06	0.00	120.4
2400	5.24	4.39	3.47	2.48	1.61	0.74	0.04	0.00	107.2
2600	5.15	4.27	3.32	2.32	1.45	0.60	0.03	0.00	96.2
2800	5.06	4.15	3.18	2.17	1.30	0.49	0.02	0.00	86.8
3000	4.97	4.04	3.05	2.03	1.17	0.40	0.01	0.00	78.8
3200	4.88	3.92	2.92	1.89	1.06	0.32	0.01	0.00	71.9
3400	4.79	3.82	2.79	1.77	0.95	0.26	0.01	0.00	65.9
3600	4.71	3.71	2.67	1.65	0.86	0.21	0.00	0.00	60.6
3800	4.62	3.61	2.56	1.55	0.77	0.17	0.00	0.00	56.0
4000	4.54	3.51	2.45	1.44	0.69	0.14	0.00	0.00	51.8
4200	4.46	3.42	2.35	1.35	0.62	0.11	0.00	0.00	48.2
4400	4.38	3.32	2.25	1.26	0.56	0.09	0.00	0.00	44.9
4600	4.30	3.23	2.15	1.18	0.51	0.07	0.00	0.00	41.9
4800	4.22	3.14	2.06	1.10	0.45	0.06	0.00	0.00	39.3
5000	4.15	3.06	1.98	1.03	0.41	0.05	0.00	0.00	36.8

[a]See Chapter 6, Table 6.2, for definition of field size classes.

[b]BOE = barrels of oil equivalent.

As with the Miocene-Pliocene trend, the estimates shown in columns 3 and 4 of Table 8.4 were used in the Arps and Roberts model to forecast the incremental future rate of discovery in the Pleistocene trend (Table 8.5). The incremental rates of discovery forecast for the first drilling increment of 200 wildcat wells were for the discovery of 25.5 fields each larger than 0.76 million barrels of oil equivalent; that is, in field size classes 9–16 (Table 8.5, row 1). This collection of fields was expected to contain 256.8 million barrels of oil equivalent, resulting in a 1.28-million-barrels-of-oil-equivalent-per-wildcat-well discovery rate during this first drilling increment. The rate of return for wildcat drilling in the Pleistocene trend was forecast to decline to approximately one-half of the level of the first drilling increment by the completion of the sixth increment when the cumulative number of wildcat wells drilled in the trend would be 2200.

The last step in our involvement with the Gulf of Mexico forecast was to present to Emil Attanasi, a USGS econometrician, the physical data on the size attributes of the geologic trends and the oil and gas fields expected to be discovered in the future, along with all the parameter estimates and incremental discovery rate forecasts. He and John Haynes of Global Marine Inc. amalgamated these data with the many different types of costs associated with finding

Figure 8.10. Number of wildcat wells projected to be drilled in the Gulf of Mexico study area by trend after December 31, 1976, as a function of marginal finding and development cost with a 15-percent discounted cash flow return assumed (Attanasi and Haynes, 1984).

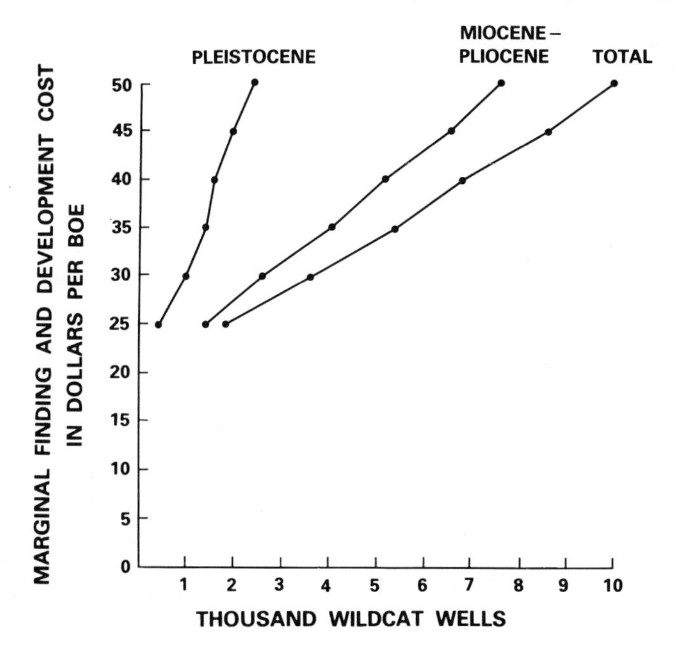

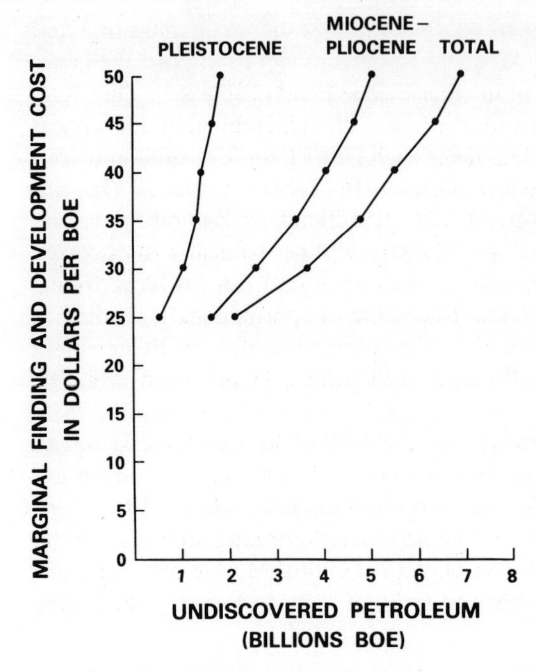

Figure 8.11. Marginal finding and development costs (in 1984 dollars) for fields to be discovered in the Gulf of Mexico study area after 1976 by trend areas and a 15-percent discounted cash flow return assumed (6000 cubic feet natural gas equals 1 barrel of oil equivalent) (Attanasi and Haynes, 1984).

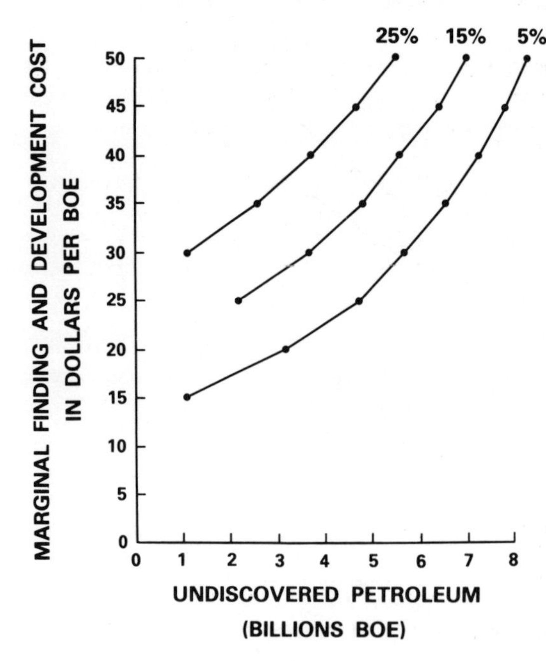

Figure 8.12. Marginal finding and development costs (in 1984 dollars) for Gulf of Mexico fields found in the study area after 1976 for 5, 15, and 25 percent required return (6000 cubic feet natural gas equals 1 barrel of oil equivalent) (Attanasi and Haynes, 1984).

and producing oil and gas. The final product from these calculations was a set of marginal cost curves (Attanasi and Haynes, 1983a, b, 1984).

Several of these marginal cost curves are shown in Figs. 8.10–8.12. In Fig. 8.10, a forecast is presented for the number of wildcat wells that would be drilled in each of the geologic trends after December 31, 1976, as a function of the marginal finding and development costs. The volumes of oil and gas that will be discovered at a 15-percent rate of return, which is conditional on the discovery rate forecasts listed in Tables 8.3 and 8.5, are shown for the two geologic trends in Fig. 8.11. From the marginal cost curves displayed in this figure, we see that if the price per barrel of oil equivalent should rise (for example, from $25 to $50 per barrel of oil equivalent), the volumes of oil and gas expected to be discovered in the Miocene-Pliocene trend will be approximately 3.5 billion barrels of oil equivalent and approximately 1.2 billion barrels of oil equivalent in the Pleistocene trend.

The affect that the internal rate of return will have on the volumes of oil and gas remaining to be discovered in the Gulf of Mexico, is shown in Fig. 8.12; for example, if we assume that the average operator requires an internal rate of return of 25 percent and that the price is set in the market at $30 per barrel of oil equivalent, we can use the graphs in Fig. 8.12 to answer the question of how much more oil and gas will become profitable to produce if the financial circumstances should change so that the average operator would require a return rate of only 5 percent. This could happen if interest rates were very high and then fell. The consequence of such a change would be that nearly 5 billion barrels of oil equivalent would move from the undiscovered category into the discovered category.

The objective of the interagency study discussed in Chapter 6 was to provide the above-mentioned series of marginal cost curves that can be used to determine how much oil and gas will be discovered under various economic conditions. In the process of producing these curves for the Permian basin and the Gulf of Mexico, we were also able to develop ideas on how to estimate the parameters of the parent population of oil and gas fields by using the concept of economic truncation. In the next chapter, we will examine how these ideas can be used to categorize undiscovered oil and gas resources.

9

Exhaustion of the Resource Base and the Future Supply of Oil and Gas

This chapter summarizes the exhaustion of the parent population of oil and gas fields by the exploration process and links it to the effect of a change in oil and gas prices to form a set of future supply categories. This concept was developed from the application of discovery process modeling, which was discussed in Chapters 6–8, and from the theory of the exploration play, which was discussed in Chapter 4.

Exploring for crude oil and natural gas within an exploration play is an economic activity that progressively attacks the parent population of undiscovered oil and gas fields that occur under the fairway of the play. The exploration process deals with large risks that arise from the uncertainties associated with the amounts of crude oil and natural gas found when a new discovery is made that starts a play, as well as with the number of wildcat wells needed to make additional discoveries in the early phase of an exploration play. The quantity of oil and gas reserves found by these successive discoveries can vary by five or six orders of magnitude, and the number of wildcat wells required to make successive discoveries can vary from 1 to more than 30 wells. Later in a play, the sizes of the fields discovered are, on average, much smaller and, therefore, less variable. As a play progresses, the proportion of wildcat wells that are successful settles down to an average of about 1 in 10.

When discovery process modelers think about the progressive character of an exploration play, we usually are mulling over how the undiscovered oil and gas fields will be exhausted by wildcat drilling over time. As a starting point, we associate this exhaustion phenomenon with the notion of drawing samples from a population *without* replacement; this means that each time a new discovery is made, there is one less field remaining to be found in the future. Any useful long-run model of the discovery process must contain a mechanism that accounts for this sampling-without-replacement characteristic.

However, just because a modeler states the sampling-without-replacement postulate, it does not mean that the structure of a discovery process model is complete. The modeler must account for the fact that even though the number of oil and gas fields remaining to be discovered is diminishing over time, long-run physical exhaustion does not necessarily control short-run rates of discovery. Aggregate discovery rate functions, such as the barrels of oil equivalent discovered per foot of wildcat drilling, can exhibit declines that, over a period of several years, have little to do with sampling without replacement. Such declines have much more to do with the formation of rising profit expectations, which, at least in part, are controlled by psychological factors in which an increase in prices of crude oil and natural gas plays a large role. These same factors were identified earlier (Fig. 4.13) as controls on the decline and recovery of the discovery success rate within an exploration play.

So an increase in the price of crude oil and natural gas can cause a surge of enthusiasm for investment in exploration ventures and a fall in the rate of discovery. When prices subsequently fall, we have a cycle. The imprint placed on the rate of discovery results in a declining rate as prices rise and a rising rate as prices decline. This type of decline occurred during the aggregate discovery rate for onshore oil and gas exploration from 1974 to 1978 (Office of Technology Assessment, 1987). The other one-half of the cycle—when prices fell and the aggregate discovery rate rose—started in early 1986. It will take several more years for the statistical information to become stable enough for comparison. However, I will wager that the aggregate discovery rate for the onshore United States will have returned to the pre-price-increase level. Why do rising prices drive aggregate discovery rates down in mature producing regions, such as the United States? The answer is that as prices spiral upward, smaller and smaller fields become increasingly profitable, but to discover one of these fields takes approximately the same number of wildcat wells or cumulative footage as it does to discover the same number of larger fields. So the discovery rate per well or per cumulative footage has to fall. The opposite happens when prices fall.

The behavior of wildcat drilling and discovery during the initial phase of an exploration play is another important topic for the discovery process modeler. During this phase, which usually covers a short interval of wildcat drilling, most of the larger fields that occur under the fairway of the play are discovered, and, as a result, a large proportion of the total oil and gas that occurs in the play is discovered. During this period, the rates of discovery are high, after which they fall quickly to a low level that is maintained for a long period of wildcat drilling. The decline in the rate of discovery from its initial high level is very fast, and only in the most unusual circumstances is it even temporarily reversed. This aspect of exploration plays was discussed at some length in Chapters 4, 6, and 7, where the observation was made that the typical profile for the aggregate discovery rate graphs for exploration plays and larger areas, such as basins, the entire United States, and the non-Communist world onshore and offshore, had a characteristic L shape (Hubbert, 1974; Root and Attanasi, 1980; Drew et al., 1983). The massive initial declines observed in the aggregate discovery rates for these larger regions, where more than one exploration play occurs, is, to my mind, nothing more than a reflection, on a larger scale, of what goes on within an individual exploration play. In these regions, the bigger exploration plays (those containing the largest fields and the largest amount of petroleum) tend to be triggered earlier than the smaller exploration plays. These large, early declines in rates of discovery, which are found in the exploration histories of region after region, whether large or small in areal extent, are caused by a second attribute of the discovery process.

The characteristic L shape that aggregate discovery rate profiles have, whether it is for an exploration play, a basin, or a larger area, is determined by a far more potent mechanism than simply sampling the parent population of oil and gas fields without replacement. Discovery process modelers have formulated a number of functions to describe the behavior of discovery process at its various stages. The all-important discovery behavior during the early stage of the play has been postulated to follow various functions of the sampling-without-replacement mechanism. This function is almost certainly more complicated than the constant power function, which was postulated by Kaufman et al. (1975). Substantial evidence suggests that the strength of this function increases as field size increases (Drew, et al., 1980, 1982; Meisner and Demirmen, 1981; Forman and Hinde, 1985). Whatever formulation of the discovery process we might choose, we have to recognize that it is progressive in the sense of sampling without replacement in the long run and a multiplicative function of sampling without replacement during the initial phase of exploration.

At this point, I would like to present a sales pitch for probability theory and its relevance to understanding the discovery process. To illustrate the lack of appreciation for the role of probabilistic mechanisms in the outcome of drilling a sequence of wildcat wells, I offer the following observation. In the community of oil and gas exploration geologists and other professionals involved in making exploration drilling decisions or in analyzing their outcome, confusion exists concerning the meaning of the most important postulate on which discovery process models are based. This postulate says that the biggest field occurring in the fairway of any exploration play tends to be discovered early in the exploration history of that play. The confusion of which I speak stems from the misunderstanding between what is said and what is heard. We modelers say the word *early,* but the word *first* is heard by almost everyone else.

Those who do not understand our methods benefit from this noncommunication because it affords them the opportunity to point out that in many exploration plays or basins or even countries, the largest field discovered to date was found on the second or third or tenth successful wildcat well. It has often been pointed out to me that this fact alone proves that there cannot be any meaningful content to discovery process modeling.

I have always found such statements curious given that it is not difficult to demonstrate that the largest fields in the fairways of most exploration plays, on average, are found very early. Sometimes, the largest field is the first discovery made in a play. I think it is nonsense to say *ex ante* that the position occupied by the largest field in the discovery sequence is anything other than the outcome of a probabilistic process. Larger fields have a higher, often very much higher, probability of being discovered early in the sequence of wildcat wells drilled in an exploration play. For that matter, the position of every discovery, large or small, in the wildcat drilling sequence can be considered to be the outcome of a probabilistic mechanism.

As seems to be commonly assumed, these mechanisms do not have pure disorder at their core. To the contrary, they usually have a great deal of causal structure. I fear that this point is easily misunderstood. In the common vernacular, the word *probability* is associated with a lack of order, a lack of cause. Maybe we should have labeled our models "causal models of the oil and gas discovery process that have well-behaved error components that make good sense if you believe in the wholesomeness of probability theory."

If it is tough for us discovery process modelers to push ideas about the order of the discovery of large fields in exploration plays, then it should be no surprise that ideas about the manner in which small fields are found will be equally difficult to communicate. By using a well-worn cliché to emphasize my point, I would be a millionaire if I had a nickel for every time I argued that small fields are found at a nearly constant rate all the way along the wildcat drilling sequence. Small oil and gas fields are discovered at this nearly constant rate mainly because they have small areal extents. We say that, in the long run, their rate of discovery is approximately modeled by applying the sampling-*with*-replacement probability model.

Before going further, we need a technical definition for "a small field in a partially explored area." A small field has a size that is equal to or smaller than the modal class of the observed field size distribution for the region in question. I hasten to point out that this definition of a small field does not yield an absolute size as compared to the rule-of-thumb definition that has existed for some time in which a small field is any field containing less than 1 million barrels of oil equivalent. Not only does this definition not yield an absolute size, it does not even yield a stable size in that the modal size class tends to migrate to the left in the observed field size distribution as an exploration play progresses. What it does do, however, is allow the historical discovery to be partitioned into two segments that can then form the necessary basis for estimating the numbers of oil and gas fields of all sizes left to be discovered and the rates at which they will be discovered in the future by using a discovery process model. The above definition for a small field is an operational definition which is directly tied to the method used to estimate the number of these fields that remain to be discovered.

According to this definition, a small field in a region with very high exploration costs, such as in the North Sea, will be any field containing less than about 100 million barrels of reserves. In most of the onshore basins in the United States, a small field would be smaller than 1 million barrels in size. In the exploration plays that have the lowest known exploration costs, a small field is approximately 100,000 barrels in size.

The key to estimating the number of fields left to be discovered and to forecasting their future rates of discovery is to use the transitional pattern found in the past rates of discovery for the fields that are larger than the mode of the observed distribution as a basis for predicting the future rates of discovery for all the sizes of fields that remain to be discovered. We refer to these fields as the "larger fields." The estimation procedure, which was discussed in Chapter 8, consists of calibrating the discovery process model against this transitional pattern of discovery rates, which is found in the larger fields, and then determining the number of fields that remain to be discovered. From the results of this forecast, a second forecast is made of the number of *small* fields remaining to be discovered. The third step is to forecast the future rates of discovery of all the field sizes.

As field size increases, the behavior of the transition pattern exhibited in the discovery rate profiles moves from linear to nonlinear. This trend toward increasing nonlinearly is caused by the replacing of the sampling-*with*-replacement mechanism, which describes the discovery rates for the small fields, with the sampling-*without*-replacement mechanism, which describes the discovery rates of the larger fields. The mathematical character of this transition in discovery rate data is the critical element that allows a discovery process model to estimate the number of oil and gas fields remaining to be discovered and the rate at which they will be discovered in the future as a function of the wildcat drilling effort.

The results produced by the discovery process models have indicated that the estimated number of small fields remaining to be discovered is very large. This is not an idle concept that needs only to be noticed before it can be set aside comfortably, as has often been done by those who argue that small oil and gas fields are of no concern because most of the oil and gas discovered in the past has been found in large fields, even if an enormous number of small fields remain to be discovered. For the discovery process modeler who is trying to predict the future by producing forecasts of the oil and gas resources remaining to be discovered in partially explored regions, estimating the future discovery rates of small and larger fields is the central issue in resource assessment. These forecasts are made to determine the size distribution of the oil and gas resources that remain to be discovered and to provide critical information for economic analyses in which supply responses to changes in economic conditions are estimated.

I have gained an appreciation for how many small fields remain to be discovered in the future from my struggle to determine how many small fields were discovered in the past. My goal has been hard to reach because troublesome operational problems shroud the effort to make this determination. Many of the small fields that were discovered before the middle 1970s were not identified as entities. Instead, the wildcat wells that discovered these fields were declared to be dry holes or the oil and gas that was discovered was simply ignored or tossed in with larger fields; occasionally, they were recorded as shows of oil and gas. As was pointed out for offshore regions, such as the Gulf of Mexico (Chapter 8), releasing previously leased blocks that had been drilled and then dropped constitutes a piece of evidence left behind in the public record which tells the tale of small fields that were discovered but not recorded.

As I accumulated examples of the ways in which the small fields were discovered and not formally recognized in the public record, whether they were misrecorded, too small to bother with individually, or ignored, I recognized that a large inventory of small fields existed and that many of these "old," previously discovered small fields would appear as "newly" recognized fields as the prices of crude oil and natural gas rose rapidly during the late 1970s and early 1980s. I could see that as prices rose, the available data from electric logs and engineering tests from the discovery wells of a large number of these "shows," dry holes, and ignored discoveries were being reexamined. The small amount of crude oil and/or natural gas which many of these reservoirs contained but was not worth recognizing as entities in the past, suddenly became economically interesting.

Some of the most numbing professional conversations that I have ever had occurred while trying to explain this twist in the recording of past discoveries. The feat yet to be performed is to convey concisely the idea that a very large number of small oil and gas fields have been found and that, although they may not contain much of the crude oil and natural gas discovered in the past, these fields are going to contain perhaps as much as 80 percent of the conventional oil and gas discovered in the future in the nonfrontier areas (today, most of the frontier areas for petroleum exploration are either under polar ice or under very deep water offshore), which, by now, are most of the places where oil and gas will be found.

By stitching together all the above information—the data that I obtained from the analysis discovery rate data within exploration plays and from larger scales, the search for an estimate of the number of small fields that had been discovered before the energy crisis and not recorded, and the performance of discovery process models—I was able to settle on an image of the physical nature of the discovery process and how it exhausted the parent population of the oil and gas fields. I based this image on the migration of density mass within the observed frequency distribution of discovered field sizes. The migration of these density masses under the assumption of a static economic regime is shown in Fig. 9.1.

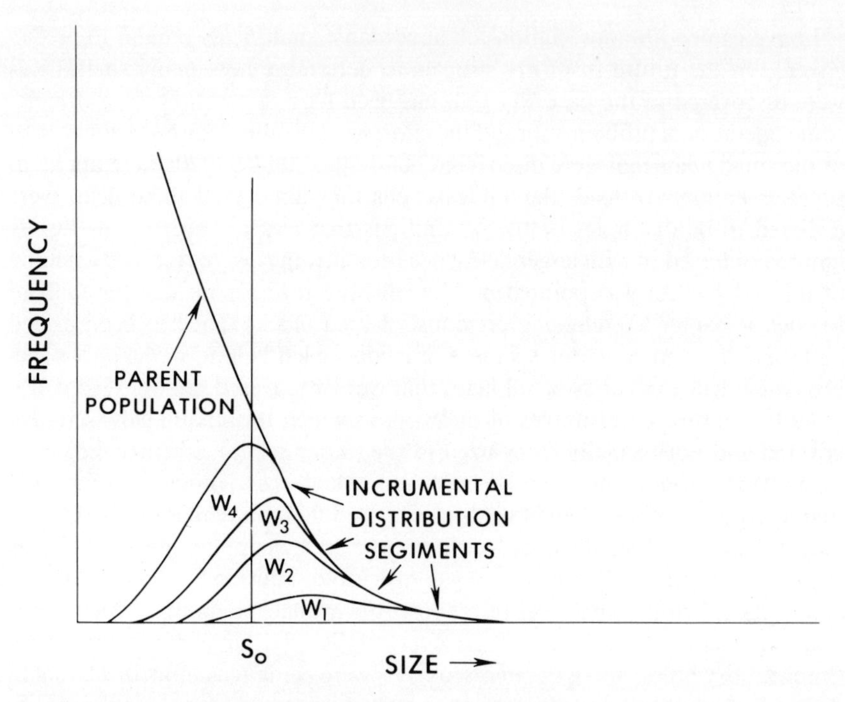

Figure 9.1. The progressive exhaustion of an oil and gas field size distribution by wildcat drilling. S_0 is point of economic truncation. W_1 to W_4 are sequential segments of the field size distribution added with W_1 to W_4 wildcat wells.

This description of the discovery process assumes that the modal field size class of the observed field size distribution is approximately the minimum size field that can be developed at a profit. This minimum field size is called the point of economic truncation for a static economic regime (price-to-cost regime). In the real world of realtime decision making, the expected flow rate of a field is the decision criterion. The size of a field as used here (that is, its ultimate productivity) usually is correlated closely with its early production rate and is far more meaningful to the analysis of the undiscovered oil and gas resources. It is defined as point S_0 in Fig. 9.1. As the level of wildcat drilling increases in an exploration play over time, say, from w_1 to w_2 to w_3 to w_4, the discoveries that will be made within each of these units of wildcat drilling will form the corresponding segments of the frequency distribution mass when collected together (Fig. 9.1). This leftward migration in the relative frequency of the sizes of observed discoveries as an exploration play matures under a static price-to-cost regime will be referred to as the effect of simple exhaustion. In the strictest sense, this definition should only be used with the parent population of oil and gas fields that occur under the fairway of a play.

In addition to the left-shifting density segments being added to the observed size distribution (Fig. 9.1) of discoveries, we also observed that within each segment (for example, w_3), some portion of the recorded discoveries (shown in Fig. 9.1) made will be smaller than S_0. On average, these discoveries are not profitable. These fields were put into production to recover part of the sunk cost and/or on the speculation that a field will prove to be larger than was initially thought. Data on the sizes of these small fields would be available because they would be developed, put into production, and, thereby, be entered into the public record. This information is collected most often by state regulatory agencies to monitor royalty and tax payments.

There is the associated collection of fields (not shown in Fig. 9.1) that have been discovered in the same size range (that is, smaller than S_0 in segment w_3) but for which no records exist. This is the collection of discoveries mentioned above that were not completed as producing fields. This collection of fields was implied to be large in number, and the assertion was made that the existence of these fields must be accounted for before a meaningful forecast of future discovery rates can be produced by using a discovery process model. If the level of economic truncation of the field size distribution could be assumed to be stationary for the foreseeable future, then the existence of these small fields can be ignored. It is reasonably certain that the price-to-cost regime will improve in the future. As it does, development of smaller and smaller field sizes will become increasingly more profitable.

We can conclude that if the prices of crude oil and natural gas had not risen as they did between the middle 1970s and the early 1980s, the evolution of exploration plays could have continued to be modeled as it had when it was assumed that only the effect of simple exhaustion was at work. If this had occurred, each exploration play would have progressed to a position of being exhausted against the relevant S_0 barrier.

The path of this progression toward simple exhaustion was significantly altered when, in the middle 1970s, prices broke upward out of the narrow trend that they had been following for many years. The nature of the response to this improvement in the price-to-cost regime at the level of the exploration play had an overall positive, but distinctly varied, theme.

Many exploration plays in the United States had evolved to the point of simple exhaustion before the prices of crude oil and natural gas began to rise during the 1970s. The collections of new discoveries made in response to these price increases had sizes that formed a density mass, the center of gravity of which is shifted to the right of the previously established points of economic truncation. In those exploration plays that had shut down under simple exhaustion before the prices began to rise, the effect of the rising prices was relatively easy to determine.

As an example of such an exploration play, I will use the Minnelusa play in the Powder River basin, Wyoming. This play started in the middle 1950s and essentially had been exhausted by the late 1960s. The bulk of the crude oil discovered in this play (85 percent through 1981) was found during the 12-year period from 1957 to 1969. Wildcat drilling peaked in 1963, when 124 wildcat wells were drilled. By 1970, the level of drilling had fallen to only 34 wells per year, and it had declined by 1971 to 25 wildcat wells (Drew et al., 1987).

The segments of the frequency distribution (Fig. 9.2) show how the observed field size distribution was first built up by a simple-exhaustion effect and later increased by the price effect. The discoveries made from the start of the play in 1957 through the end of 1970 are attributed to the effect of simple exhaustion, which was active under the static price-to-cost regime that existed during this time period. The second segment of density mass shown is a transition segment (discoveries made between 1971 and 1975), which is attributed to a mixture of discoveries composed of a few fields that would have been discovered under the pre-1973 price-to-cost regime and to those discoveries made because the price of crude oil was increasing. The third segment (1976 to 1981) contains a collection of fields whose discovery is almost entirely attributable to the price effect. Nearly all the fields in this third segment contain less than 1.52 million barrels of crude oil. This segment is clearly shifted to the left of the previously added segment. Recognizing that a certain volume of crude oil and natural gas is left behind in such situations, I will identify the group of exploration plays that have shut down against a cost barrier (S_0 in Fig. 9.1) as belonging to a future supply category of oil and gas (Table 9.1, category I).

The effect of the price increase during the middle 1970s to early 1980s on

Figure 9.2. Migration of frequency distribution mass in the Minnelusa exploration play, Powder River basin, Wyoming.

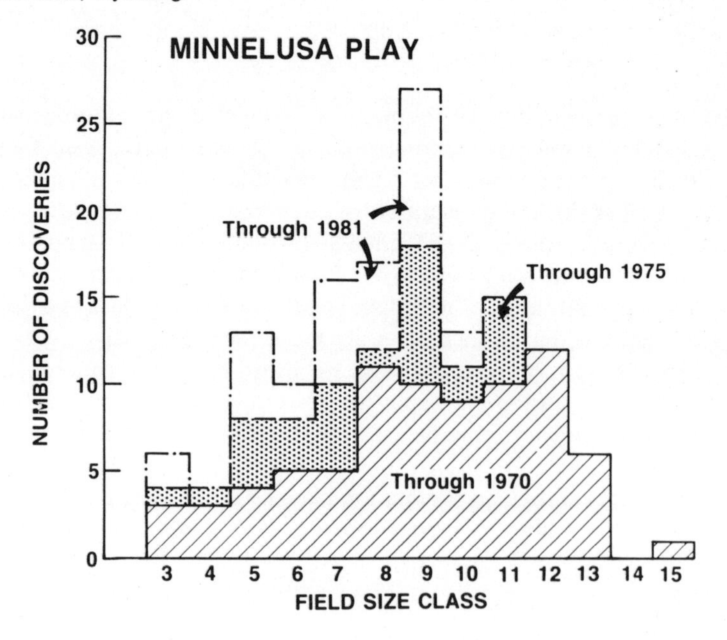

Table 9.1. Definitions of Future Oil and Gas Supply Categories by Exploration Play Status Defined for a Given Point in Time

Future supply category	Price increase needed for cycle of future discovery	Status of play before price increase	Example
I	Yes	Shut down after cycle[a]	Minnelusa
II[b]	Yes	Ongoing	Miocene-Pliocene
III	Yes	Identified but drilling cost too high	Pleistocene
IV	Yes	Well documented by geologic studies	Deep-water offshore Atlantic reef
V[c]	Yes/No	Pure speculation	Many examples, such as Southwestern overthrust
VI	No	Ongoing	Miocene-Pliocene
VII	No	Not yet recognized but would be profitable	?

[a]A play cycle is complete after shutting down against cost barrier S_0.

[b]Companion category to category VI.

[c]The willingness to take extremely large risks for extremely large gains can drive pure speculation in nearly any price-to-cost regime.

Note: Below dashed line no price increase needed.

those ongoing exploration plays is not easy to discern because the discoveries that would have been made under the previous price-to-cost regime are mixed in with the additional discoveries caused by the price increase. The operational problem of isolating these two effects (the simple-exhaustion effect versus the price effect) also is complicated by the normal variability that exists in the individual field size discovery profiles. This variability is large enough that even when the experienced analyst is attempting to isolate the simple-exhaustion effect, it can be concluded that a cumulative discovery profile has reached a true asymptote when it is stuck on a false asymptote for an abnormally long interval of wildcat drilling. I recognize that the variability that exists in discovery rate data should be discussed, but I am leaving it for another time as I plunge past the issue to continue the definition of a number of additional future oil and gas supply categories (Table 9.1).

The separation of an ongoing simple-exhaustion effect from the price effect is more of an issue in data analysis than it is in the industrial organization of the exploration business. The nub of this issue relates to the construction of a rationale to separate the discoveries that would have been made in an ongoing play had the price increase not occurred from those that the price increase generated. I have chosen to follow an empirical path in the construction of this rationale.

The classification of future oil and gas supply categories shown in Table 9.1 relies, in part, on the application of this empirical rationale. Note first that when an ongoing exploration play is subjected to the stimulus of rising prices, the play contributes to two future supply categories. The first of these contributions is a result of the price-to-cost regime under which the play started (Table 9.1, category VI), and the second contribution is induced by a price increase (Table 9.1, category II). The isolation of the contributions to these two categories is illustrated by using the Miocene-Pliocene play of the offshore Gulf of Mexico (Fig. 7.2).

The analysis of how this exploration play contributed to these future supply categories starts with an inspection of the observed field size distribution over the history of the play. The temporal migration of the frequency mass in the observed field size distribution for the Miocene-Pliocene trend in the offshore Gulf of Mexico is shown in Fig. 9.3. The uppermost frequency distribution in this figure shows the observed distribution in the combined trend in 1960, when approximately 650 wildcat wells had been drilled. The frequency distribution in the middle of this figure shows how the observed distribution has grown and shifted after an additional 1750 wildcat wells were drilled between 1961 and 1970. During this time period, the exploration price-to-cost regime was nearly static. Minor fluctuations did occur during this period, but the big changes in prices and costs came after this period; for example, the majority of the changes in posted crude oil and natural gas prices came after 1973, when they rose tenfold by the early 1980s. Given a nearly static price-to-cost regime between 1960 and 1970, we can attribute the growth in frequency distribution mass from the top distribution to the middle distribution (Fig. 9.3) to the effect of simple exhaustion; that is, the observed migration to the left in frequency mass is not related to a change in the price of either crude oil or natural gas. During this period, fields were being found that were profitable to discover, even though they were, on average, becoming smaller and smaller as time progressed.

From 1960 through 1970 the sampling-without-replacement effect for the larger fields was obviously at work; no additional fields were being discovered in field size classes 18 or 19 (Table 9.2). These larger fields had been discovered by the time 525 wildcat wells had been drilled in the combined trend. Discoveries in the next smallest size class (size class 17) still were being made; five discoveries were made between 1960 and 1970. In each successively smaller size class, the number of discoveries made during this time varied but clearly tended to increase going downward toward smaller and smaller field sizes (Table 9.2). This growth in the relative frequency is attributed to the simple-exhaustion effect.

As the price increased during the middle 1970s, the observed frequency distribution grew (Figs. 9.3 and 9.4, Table 9.3). This growth in the observed frequency distribution is attributed to a mixture of the simple-exhaustion and the price effects. It may be more meaningful to describe this mixture as the addi-

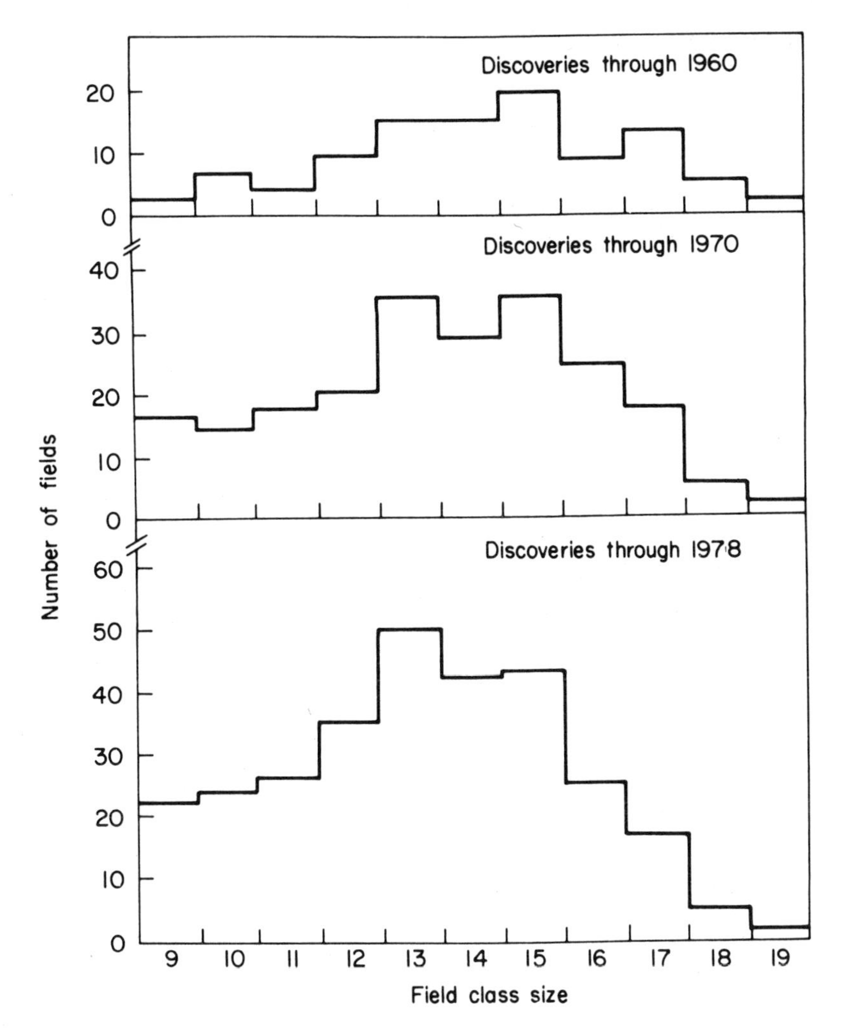

Figure 9.3. Observed field size distributions at the points in time for the combined Miocene and Pliocene trend in the Gulf of Mexico offshore (Schuenemeyer and Attanasi, 1984).

tion of the price effect to an ongoing exhaustion effect. From 1971 to 1978, the price of natural gas rose from approximately $0.20 per thousand cubic feet to between $1.00 and $1.25 per thousand cubic feet, and the price of crude oil rose from around $3.50 per barrel to approximately $20.00 per barrel. Much of this price rise (posted prices) came in the second one-half of this period. Prices continued to rise thereafter into the early 1980s, until the price of natural gas was over $5.00 per thousand cubic feet, and crude oil was over $30.00 per barrel.

Defining a foolproof analytical scheme to separate the collection of discov-

Table 9.2. Simple Exhaustion in Combined Miocene and Pliocene Trend Using Number of Discoveries 1960–1970

Size class[a]	Discoveries made through[b] 1960	1970	Increase	Proportional increase
19	2	2	0	0
18	6	6	0	0
17	13	18	5	0.39
16	8	24	16	2.0
15	18	36	18	1.0
14	15	29	14	0.93
13	15	35	20	1.33
12	9	20	11	1.22
11	4	18	14	3.50
10	7	16	9	1.29
9	3	16	13	1.29
Total	100	220	120	1.2

[a]See Chapter 6, Table 6.2, for definition of size classes.
[b]Drew et al. (1982).

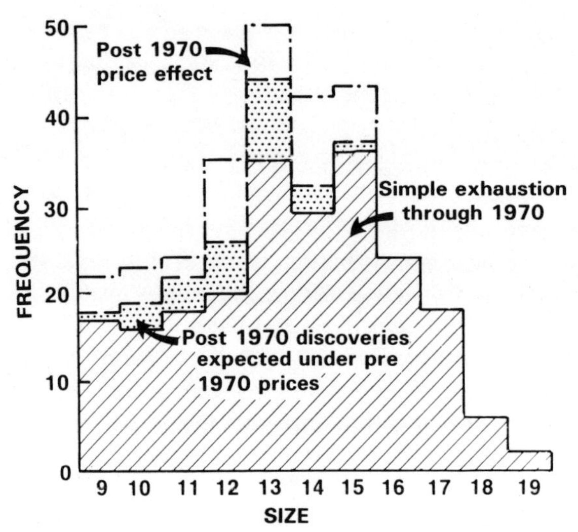

Figure 9.4. Migration of frequency distribution mass in the combined Miocene and Pliocene trend, Gulf of Mexico offshore.

Table 9.3. A Mixed Price and Simple Exhaustion Effect in Combined Miocene and PlioceneTrend Using Number of Discoveries 1971–1978

	Discoveries made through[b]			
Size class[a]	1971	1978	Increase	Proportional increase
19	2	2	0	0
18	6	6	0	0
17	18	18	0	0
16	24	24	0	0
15	36	43	7	0.19
14	29	42	13	0.45
13	35	50	15	0.43
12	20	35	15	0.75
11	18	24	6	0.33
10	16	23	7	0.44
9	17	22	5	0.29
Total	221	289	68	2.88

[a]See Chapter 6, Table 6.2, for definition of size classes.
[b]Drew et al. (1982).

eries made during any time period into respective subsets of discoveries attributable to simple-exhaustion and the price effects is not as straightforward as it may seem. Of the several difficulties to be overcome in this analysis, the most demanding is explaining how exploration operators respond to a price increase. The analysis of historical time series data may not reveal whether or to what degree the exploration operators acted before prices rose or waited for prices to change and then responded. The variability in the oil and gas discovery rate data mentioned above is an added menace in this type of analysis. Lagging the price series can help to isolate a price effect when the response variable is univariate, but when the response variable is the relative growth in the mass of a frequency distribution, the analysis is more difficult. My preference in these sorts of determinations is to explain which mechanisms are at work from an industrial organization viewpoint and then to reckon their magnitudes by using a collection of industrial information intertwined with graphic displays of data followed by formal data analysis.

The following story shows how industrial information was used to explain some of what was going on with price expectations for natural gas in the Gulf of Mexico offshore during December 1970. One morning, I paid a visit to the exploration and production planning department of the oil company that employed me. When I arrived at the department, I could see that a most unusual event must have occurred. The entire staff was standing in the lobby surrounded by computer output. The normal routine in this department was one of extreme order and decorum. Desktops were neat and tidy except for a pad of paper and a book or two. There was never any noise, and no coffee cups were ever in sight except for the 10 minutes or so after the coffee man passed through the outer hallway. No one spent any time in the corridors. So whatever was happening was a significant event. I was told that our company had done very poorly in the lease sale in the Gulf of Mexico the day before and that several of the staff had been up all night making computer runs to try to figure out why. Tom Campbell, the fellow I had come to see, said, "We came up short in the sale, and we are going to have to explain to New York why we blew it!"

The conclusion reached a few minutes before I arrived was summarized in a table that had been stuck up on the wall in the lobby. The staff, which was huddled around this piece of paper, was pointing at different groups of companies and what prices they had used yesterday in their bids. Not knowing any better, I asked how our staff could obtain such price information. I thought it would be a closely guarded secret within each company. Tom Campbell chuckled and explained that the prices for each company were estimated by assuming that each operator had access to about the same quality and quantity of geologic and geophysical information. The wisdom of the exploration staffs also was assumed to be about the same. By using the bid data and these assumptions, they ran the bidding model backward to estimate the prices used by each company.

At the bottom of the list were the major integrated companies and their combines that had bids based on natural gas prices of $0.24–$0.30 per thousand cubic feet. At the top were the gas pipeline companies and their combines and their bids that ranged up to $0.38 per thousand cubic feet. We did not pick up much with our bids based on mid-$0.20's per thousand cubic feet prices. Tom Campbell explained that it was a whole new ball game in the Gulf of Mexico. He concluded that the gas transmission companies must be betting that prices would go up by at least 50 percent. Before the morning was over, the head of the planning department appeared and announced that our chief executive officer had just shipped $10 million to one of the pipeline companies for 10-percent ownership in the leases it had obtained yesterday. Our exploration office in Houston was furious, but the chief executive officer had said that if they were not able to keep our company in the oil and gas business, he would!

Based on the above anecdote, I must argue that, by the end of 1970, the exploration operators were bidding up the price of natural gas in the Gulf of Mexico. Because of the nebulous and quicksilver nature of bidding and price increases, the effective price was not the price negotiated in the present sales contract or in the next one or even the tenth one. No records will ever be published that will reveal what the effective price of natural gas really was the morning after that December 1970 lease sale. We were looking at a situation in which the price of natural gas was moving upward well before the price increases caused by the oil embargo of 1973 occurred.

All we could say on that day in December was that decisions were being made that assumed higher prices would be coming along in the near future.

How, then, can we ever perform a direct microscale analysis without being on the scene when bids are prepared for the next lease sale or when acreage or reserves are being sold? We must realize that if we are too demanding, the use of the direct data analysis approach to estimate the effect of a price movement on a response variable, such as a change in a rate of discovery of oil and gas fields, may well elude us. I always wonder when I take a price series out of one of those softbound tomes published by a professional association or regulatory commission whether such series are lifeless concoctions that contain little information on the prices actually used at the time to make the decision of whether to buy acreage or to drill wildcat wells. Parallel to this question about effective prices is the issue of costs. Estimating costs before the fact, of course, can be as elusive as determining the effective price used in a transaction. My intuition tells me that the available exploration cost data have a higher information content than the available price data.

Debating the issue of whether a recorded cost series means more than a recorded price series is good entertainment, but we have to get past that point and decide how we will extract information from the data on hand. Left to my own devices, I usually choose to look at how a response variable changes after a price increase or decrease has begun to take effect and then to watch it for a while. Although simple, this analytical scheme may be as near as we can come to a reckoning at the microeconomic level of the effect of a price change on the rate of discovery.

Having explained my approach to the analysis of discovery rate data, we are now ready to look at a device useful in separating the contributions to future supply categories VI and II (simple-exhaustion versus price effects) for the Miocene-Pliocene play. The most certain diagnosis of what has occurred during the execution of a play is gained from an examination of the incremental and/or cumulative discovery graphs for each field size class (see Figs. 7.4–7.9, 7.13–7.21). Elements of this diagnosis are shown below by using the data from Figs. 7.17–7.19. Inspection of these graphs shows that, around 1970, the rates of discovery for some of the field size classes had begun to accelerate. This upward break in the discovery rate is particularly obvious in field size classes 15 (Fig. 7.17) and 12–14 (Fig. 7.18). The cumulative discovery graphs for the smaller fields (0.76 million to 6.07 million barrels of oil equivalent) also exhibit some increase in the rate of discovery (Fig. 7.19) at about this same time. These upward breaks in slope show that gas fields in size classes 12 through 15 in particular have become economic targets as a result of the price increases of the 1970s. In a few words, the success rate increased because these fields were recorded as discoveries when they were discovered.

The observed distribution of discoveries in this trend had been truncated *before* 1970 in the vicinity of field size class 14 (24.3 million to 48.6 million barrels of oil equivalent; Figs. 9.3 and 9.4) by the high cost of exploration in the trend, thereby leaving reasonably large fields undiscovered. As a consequence of the relatively large sizes of the oil and gas fields below the truncation point, it should be expected that the sharp rise in prices would result in the transference of a relatively large volume of crude oil and natural gas from the inventory of undiscovered fields into the producible reserves. An important generalization can be made from this observation. When a resource assessment is to be made for two similar exploration plays, the size of the assessed inventory of undiscovered resources should be larger for those plays that have high historical exploration costs.

After 1970, the device used to separate the simple-exhaustion and the price effects in the Miocene-Pliocene play was an extrapolation hooked to the upward break points in the cumulative discovery rate graphs for each field size class. The application of this extrapolation was motivated by recognizing that the price increases rejuvenated the exploration process in the trend. This rejuvenation caused the discovery process to penetrate further into the parent population of undiscovered oil and gas fields. This penetration was focused in the vicinity of the previously established mode of the observed distribution. The extrapolation device estimated the number of fields that would have been discovered had prices not increased; that is, it was used to estimate the number of fields in size classes 15 down through 9 that would have been discovered if the price-to-cost regime of the 1960s had persisted through the 1971–1978 period. The number of fields discovered through 1978 in the combined trend that was attributed to the effect of rising prices was calculated to be 41 fields (Fig. 9.4); this is out of a total of 68 fields discovered during this time period (Table 9.3). The total volume of crude oil and natural gas contained in these 41 fields is estimated to be 978.3 million barrels of oil equivalent. Between 1971 and 1978, the progressive leftward shift of the mode of the observed field size distribution from class size 15 to class size 13 resulted principally from increasing prices.

Had prices decreased instead of increased, this same scheme could have been used to make an estimate of the number of fields that were *not* discovered and that would have been had the price-to-cost regime remained static. The diagnostic elements in this determination would be an abnormally abrupt decline in wildcat drilling and a downward break in the recorded rates of the discovery of the smaller fields; the rates of discovery of intermediate fields would continue along as a function of wildcat drilling at about the expected rate.

If the procedure presented above is reasonable for isolating the effects of simple exhaustion from a price increase within an ongoing play, then we must ask ourselves if we can perform similar analyses for exploration plays associated with different prices and costs; for example, can we find evidence that a price increase has caused an exploration play in a later time to go through a cycle of wildcat drilling and discovery and shut down against a cost barrier (S_0 in Fig. 9.1) when previous costs were too high for the play to be pursued profitably? It is assumed that such a play had been shown to exist by the drilling of a few wildcat wells that resulted in a few discoveries. The answer is, of course, in the affirmative. Exploration plays in this condition are defined as belonging to future supply category III (Table 9.1). The Pleistocene exploration play of the

Gulf of Mexico offshore is an example of an exploration play that belonged to this future supply category during the middle 1960s through early 1970s.

The location of the Pleistocene trend is shown in Fig. 7.2. The area of this fairway is estimated to be about 33,600 square kilometers. In a sense, the exploration play that transpired in this trend was a simple extension of the play that had been going on throughout the 1950s and the 1960s in the Miocene and the Pliocene trends, which are directly adjacent. The location of many of the structures that trapped the hydrocarbons in this trend were mapped as an extension of the ongoing exploration in the two older trends.

Enthusiasm for exploration in this gas-prone trend had been suppressed throughout the 1960s by the higher cost associated with deeper water, lack of pipeline infrastructure, and, above all, the low price of natural gas. It is difficult to pinpoint the exact time when it was obvious that the reservoirs in the Pleistocene trend contained mostly natural gas; this made them unprofitable exploration targets because of the low price of natural gas. From the anecdotal information available, it was sometime in the early 1960s.

From my own experience in the exploration business, I feel that I can assert that leasing and drilling decisions were not based on the expectations of increasing natural gas prices until the late 1960s. It would then be reasonable to conclude that the 65 wildcat wells drilled and the nine discoveries made in this trend before 1966 were made when prices and costs were not expected to change as dramatically as they did (Table 9.4). Between 1966 and 1970, an additional 123 wildcat wells were drilled in this Pleistocene trend, which resulted in only five additional recorded discoveries. Obviously, far more than

Table 9.4. Effect on an Exploration Play (Pleistocene Play) that Shut Down in the 1960s and Rejuvenated after 1970 When the Price of Natural Gas Increased

	Incremental discoveries[b]		
Size class[a]	Through 1965	1966–1970	1971–1976
18	0	0	1
17	1	0	2
16	0	0	8
15	2	0	7
14	4	1	21
13	2	0	19
12	0	2	15
11	0	1	19
10	0	0	11
9	0	1	8
Total	9	5	111
Incremental wildcat wells	65	123	680

[a]See Chapter 6, Table 6.2, for definition of size classes.
[b]Drew et al. (1982).

five fields were discovered during this time period. I suspect that many of the wildcat wells that were being drilled were being put in place to test the idea that crude oil might occur along with natural gas in the very attractive structural targets in the trend.

This lack of interest in going after natural gas soon changed because the nation's store of natural gas reserves had been dwindling from a position of extreme surplus in the early 1950s to one of shortage in the early 1970s. A lack of foresight on the part of regulatory commissions often is cited as the cause of the creation of this shortage. For our purposes, it is important only to note that the price of natural gas had been held at a very low level. When the price started to rise, the large volume of natural gas that had been discovered and set aside or avoided in the various trends in which it had been identified was ready and waiting to be converted into reserves. Although this conversion would take a fair amount of wildcat drilling effort to identify its exact location, no great risks would be involved.

My view of the situation in many of the gas-prone regions of the United States in the late 1960s is based on the notion that for more than 50 years, we had been searching mainly for crude oil in the United States. In the process, we had drilled into a lot of natural gas reservoirs, and, as pipeline connections and quotas became available after World War II, many of the larger gas discoveries were completed and put into production. At the same time, many were left underdeveloped, and a number of the gas-prone trends that had been identified were ignored.

To prove the point about how little value natural gas had in the United States, we need only look at how much of it was flared in the production of crude oil before the antiflare regulations of the late 1960s—perhaps as much as 40 trillion cubic feet. The flaring of natural gas has come to be viewed by many as the most wasteful use of a natural resource in the history of mankind. I remember asking a sage of the oil business what he thought was wrong with natural gas that it had been treated so carelessly. I observed that it was a clean, efficient fuel that we covet today. His answer was that its only sin was that it could not be carried in a bucket.

As the 1940s turned into the 1950s and the 1950s turned into the 1960s, a situation that had been created in the natural gas business in which a price increase that put the value of natural gas on a par with crude oil was going to make the discovery of natural gas, at least for a while, much easier. Column 4 in Table 9.4 shows what happened in the Pleistocene trend of the Gulf of Mexico offshore when the price of natural gas rose in the 1970s. In the period between 1971 and the end of 1976, the oil and gas industry drilled an explosive burst of 680 wildcat wells in the Pleistocene trend. This is nearly four times the number of wildcat wells that had been drilled previously. With the drilling of these wildcat wells, 111 fields were discovered. Except for the discovery of one large oil field, the Eugene Island Block 330, the vast majority of the reserves contained in these fields proved to be natural gas, as was expected.

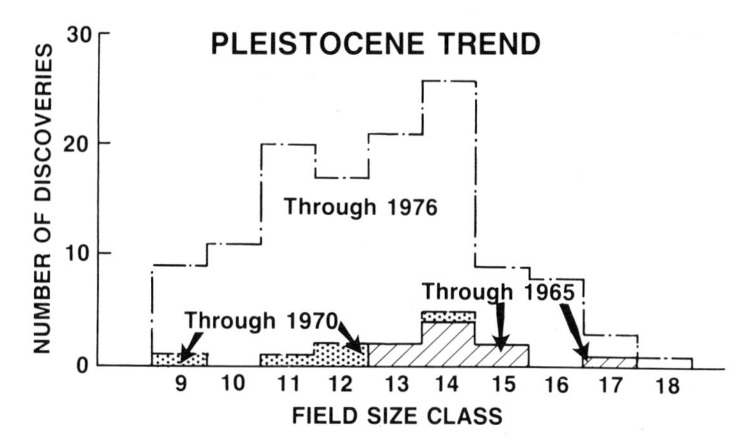

Figure 9.5. Migration of frequency distribution mass in the Pleistocene trend, Gulf of Mexico offshore.

The growth of the observed frequency distribution of discoveries for the same time intervals used in Table 9.4 is displayed in graphic form in Fig. 9.5. This figure shows how dramatic the growth in discoveries in this exploration play was after being stimulated by a massive price increase. The aggregate volume of reserves contained in the 111 fields discovered during the six-year period is approximately 3.8 billion barrels of oil equivalent, of which more than 3 billion barrels of oil equivalent (18 trillion cubic feet) is natural gas. The conversion of this volume of natural gas into producible reserves is, to my mind, totally attributable to the combination of the expectation and later realization that natural gas prices were going to increase (Table 9.1, category III). Future supply category III contains all those exploration plays that, after having been identified by drilling, drift into a state of quiescence because the fields that occur are, for the most part, smaller than the economic cutoff size of S_0 (Fig. 9.1).

The identification of additional future supply categories can be expanded well beyond the four set out above. In the exploration business, the circumstances that contribute to the creation of such categories are nearly endless, but I will limit myself to adding only an additional three, which will bring the total to seven (Table 9.1).

The first of these new categories contains the type of play that requires a very large price increase to go through a simple-exhaustion cycle (Table 9.1, category IV). The second category can contribute future supplies either with or without a price increase (Table 9.1, category V). The third category deals with the situation in which new exploration plays will be uncovered in the future solely as a function of the fact that the science of geology will progress (Table 9.1, category VII). The number of these plays left to be found in the clastic wedges of the world is uncertain.

The exploration plays included in category IV are those that have fairways thought to be favorable from the study of their geology and/or geophysics and perhaps of the results of the drilling of a stratigraphic test well. An example in this category is an exploration play that was identified about 10 years ago and would not be recognized today had the oil embargo in 1973 not occurred. It is the deep-water Atlantic Reef play off the east coast of the United States. Through a draftsman's use of a yellow roll of stick-on line marker on a poster prepared for a press conference, this play became widely known as the "Yellow Reef play." Nobody has been able to convince me that we would have a Yellow Reef play today if the price of crude oil was still $3.00 per barrel. Very high prices will be required for this play to go through a simple-exhaustion cycle, given that crude oil and/or natural gas occurs. How high a price? My guess is it would be beyond the $100-per-barrel level.

Future supply category V is a personal favorite of mine. This category, which includes a source of future supply that is difficult to quantify in purely rational terms, provides for situations in which new fields are discovered as a result of the decisions to drill wildcat wells that are not justified by using the crutch of comparative risk calculations; rather, the elements of risk taking and the rational profit expectation based on the study of comparative risks square off antagonistically. To explore for crude oil and natural gas in this category, we must be willing to overpower the conventional managerial wisdom taught in our business schools. When an exploration operator decides where he wants to punch a well down, no "bean counter" is going to turn his head by using a negative present value calculation for the project based on a comparative risk calculation. By controlling the flow of data into these calculations, the exploration operator can effectively jiggle most of the results that come out of such calculations.

Another reason for identifying this fifth category in a classification of future supply possibilities is to account for the surprise discoveries that are always occurring in the oil and gas exploration business. Why not recognize before the fact that crude oil and natural gas are going to be discovered in places that, according to geologic studies, are not considered likely to contain commercial accumulations? There are going to be discoveries made that are the result of the decisions of misguided operators who drill wildcat wells where no rational person would have done at the time. In our attempt to forecast the future discovery of crude oil and natural gas, we should allow for oil and gas coming from such a source as category V before the fact, rather than chasing along behind such events trying to explain why they occurred. For myself, I see exploration operators who work in category V as being out on the frontier—leaders, if you will—taking abnormal risks to make discoveries.

It is an oddity of the exploration business that the reward for the willingness to take such risks often goes to someone else. This occurs because of the fluidity of the information gained by one operator as a result of his willingness to take risks. It is a truism that information moves from one operator to another very quickly; as a result, it is impossible to keep ideas and data in the earth sciences secret for very long. Consequently, an exploration play uncovered in category V as the result of a wild idea can quickly become a well-established fact and then can move out of this category and into category II, III, or IV (Table 9.1). The impact that a price increase has on stimulating exploration in category V is linked to a heightened ability to bear risks rather than the willingness to take risks. It is my belief that as prices increase, the ability to take risks rises more from an increase in the flow of available risk capital into the exploration sector of the economy from secondary sources than as a response to the rising expectation of the future long-run monetary rewards.

The last future supply category, VII (Table 9.1), is intended to include those exploration plays that, when uncovered, would have been profitable under the current or even less favorable price and costs. If and when these plays are uncovered, they would go through an exploration play cycle and shut down against the existing cost barrier S_0 (Fig. 9.1). I assert that these types of exploration plays will be uncovered in the future as risks are taken to try out the new ideas developed to explain where petroleum occurs. On the way to being proved profitable, these plays may temporally pass through the grip of pure speculation (category V). On the basis of percentages, the majority of the exploration plays that have been identified in the past began as the product of some mixture of new geologic ideas and pure speculation. As yet, the origin of very few exploration plays can be attributed to upward price movements. In the future, these types of plays will be uncovered more frequently.

I fully recognize that making a list of categories of future crude oil and natural gas discoveries is, for many reasons, not a very tidy business. One could be defeated from the start by knowing that the boundaries of any such set of categories have to be fuzzy. There is also the concern over whether a change in price can really be shown to trigger an event that causes a new idea to be generated, which, when pursued, points the way to a new untested place where crude oil and natural gas occur. Certainly, as prices rise, this must be possible.

As my training in microeconomics requires, I try always to keep in mind that current price levels support current exploration activity and that changes in price levels will change activity levels, thereby keeping me under the influence of the classic price-quantity calculus of this theory, but, and this is a big "but," the risk elements inherent in the exploration business cause the available population of exploration ventures to be a rather disjointed collection. To add to this stewpot of a real-world situation, we have the nonuniform perception of risk, which, in turn, is mixed with varying degrees of willingness to accept risks. From this, I argue that some crude oil and natural gas will continue to be discovered even if the wellhead prices of both commodities should fall to very low levels. The driving force is that men and women, either as individuals or as members of a corporate entity, like to gamble. This propensity will be supported by the venture capital that is always available whether economic conditions be boom or bust. It does not matter much where this type of capital comes from—equity, tax shelters, or subsidies—or who provides it.

As always, the existence of risk and the willingness to take it in all its many forms clouds the relation between what would have occurred sooner or later under an existing price-to-cost regime and what a price change causes to happen. It is then impossible, in a deterministic sense, to know which oil and gas prospects would have been drilled or not drilled under an existing price-to-cost regime had a price increase or decrease occurred. There always will be a level of subjectivity required in making these determinations. The best we can do is to erect a system of categories that has admittedly fuzzy boundaries and that we can use as the basis of a language to analyze the subject matter that makes up the fields of oil- and gas-resource assessment. As we proceed, we will improve these categories by looking at the subfields of this discipline, which include the degree of physical and economic exhaustion of the resource base, the size distribution of oil and gas fields, and the rates at which different size gas fields will be discovered in the future.

In conclusion, the seven future supply categories mentioned above do not cover what normally are termed "unconventional sources." These sources of crude oil and natural gas usually are thought to include very heavy oils, tar sands, oil shale, liquid and gaseous hydrocarbons converted from coal, very tight gas sands, and gas from fractured shales and coal seams. Because of engineering and cost considerations, the future supply of oil and gas from these sources as a function of future price expectations usually is treated separately. To be sure, someday these will become important sources of oil and gas. The main reason they are not able to compete very effectively with the conventional sources is that we are not willing to allocate the large subsidies required to make production from these sources profitable. When the conventional stocks of oil and gas have dwindled to where the prices go up to and stay up at $100–$200 per barrel of oil equivalent (current dollars), then, and only then, will we see the new age where hydrocarbons are produced in large volumes from unconventional sources. For the present and the intermediate future, at least, conventional sources will continue to provide most of our oil and gas. This is the way it is going to be. No process has been developed that can produce crude oil and natural gas as cheaply as that which flows unaided out of conventional reservoirs, into well bores, and up to the surface. The valves on the christmas trees of the wells in the world's largest fields are going to control our supply of energy for some time to come, whereas the small fields that were discussed in this chapter are going to provide most of the crude oil and natural gas that will be discovered in the future.

10

Quantitative Mineral-Resource Assessment

In March 1982, I became chief of the Branch of Resource Analysis in the Office of Mineral Resources at the USGS. Early in my tour as branch chief, which is a rotational administrative position, I was told by Alexander Thomas Ovenshine, the chief of the Office of Mineral Resources, to steer the branch toward a specific objective—to develop quantitative methods of performing mineral-resource assessments on the various packages and tracts of land owned by the federal government. He believed that we should develop procedures to estimate the number of undiscovered mineral deposits as we had done with the oil and gas fields. He saw numerical assessments as a big step toward producing meaningful summaries of the quantity and quality of undiscovered mineral resources.

Ovenshine felt that some of the progress we had made in the last decade in discovery process modeling and the estimation of undiscovered oil and gas resources by using field size distributions could be transferred to mineral-resource assessment. If it could not be transferred directly, then surely something could be done to move ahead on the issue of how to present more meaningful conclusions of our assessment of federal lands, particularly wilderness lands. It was clear to me that the order of the day was to move our assessment methods as quickly as possible toward an explicit numerical form. I think Ovenshine's main desire was to present one of our most important products in a form that was free from our complicated geologic terminology. Because he wanted us to express ourselves clearly to the nongeologist, he believed that simple numerical tables of data listing estimates of the number of undiscovered deposits and the amount of the various metals they contained was how this could best be done.

Personnel in the Office of Mineral Resources had been working on the assessment of undiscovered mineral resources occurring on federal land for a very long time and on the evaluation of wilderness lands since the mid-1960s. It was true that we had written our reports to the public as geologists and geophysicists wrote for each other; that is, we had chosen to write narrative scientific reports that were acceptable among ourselves. Usually, all we said was that there was a high or medium or low potential for the occurrence of undiscovered minerals in the land tracts that had been evaluated.

The object of Ovenshine's quest went far beyond presenting these sorts of vague qualitative conclusions. As he watched the interaction of the political forces at work in the country during the 1970s, he became concerned about the estimates and the books that we were producing for the public. He wanted to be responsive to the need to estimate the economic value of the mineral production lost from the mining sector by land withdrawals in the wilderness program. In addition, a scarcity of cobalt led to the more focused question of how much cobalt and certain other metals, which were classified as critical and strategic commodities, remained to be discovered within federally owned lands. As I recollect, the questions that arose out of the cobalt shortage were seen by Ovenshine as just another reason for the Office of Mineral Resources to develop methods of analysis and presentation of mineral-resource assessments that were understandable to the diverse group of people that it considered to be its clients.

Ovenshine was not alone in his quest because, by the early 1980s, many of the geologists at the USGS had awakened to the same idea. We realized that our clients wanted our assessment reports to contain tables showing our estimates of the tons of producible metals contained in the yet-to-be-discovered mineral deposits that occurred in each wilderness area, national forest, or other region that we had been charged to investigate. Although we sensed the desire for numerical assessments, we as a group felt uneasy about performing such assessments because doing so required the application of probability theory. Even though it was becoming common knowledge that numerical assessments were useful for communicating results and that they were desired by downstream users, we still had trouble making statements about very uncertain events, such as the probability that a given type of mineral deposit occurred in a given tract of land. Even today, some geologists remain firmly rooted in the deterministic tradition and refuse to generate even qualitative probabilistic statements about the undiscovered mineral endowment of a land tract.

This resistance was not limited to the economic geologists and geophysicists who worked up geologic and geophysical data on a land tract. Those of us who had experience in oil- and gas-resource assessment and were used to working with the numerical end of the assessment business were slow to jump into the middle of the debate and to advocate the widespread use of numerical representations of undiscovered mineral-resource endowments. We knew from previous experience that the process of attempting to place numbers where there had only been a concept was going to be traumatic. Most of the ructions that I saw as Ovenshine began his program were caused by our geologic colleagues who announced loudly and clearly that numerical mineral-resource assessments had no basis in proper geologic reasoning. There were also problems with the idea of finding out exactly who the audience was for mineral-resource assessments and how they wanted to use them. I thought it might be useful to go around the nation's capital to find these people in the flesh so that we could ask them what format and assessment would best serve their needs.

I soon discovered that to seek out these clients and to ask them what they needed in a mineral-resource assessment and then to make a tidy summary would be a difficult, if not an impossible, task for any single group to achieve. I learned that there were many more facets in the mineral-resource assessment process and its products than I had expected. I learned that when we became involved with the wilderness program, we had become involved with the issue of the multiple uses of federal lands, and, by the time I became branch chief, the mineral-resource assessment issue regarding federal lands had a substantial history. There were several polar positions and vortexes clearly evident within the mass of subject matter and opinion that made up the field. I was to find out first that, during the previous two decades, the environmentalists had fairly won the day against the promineral-prodevelopment folks. I do not think that anyone disagreed with the position of the environmentalists. It had been realized beyond any reasonable doubt that mankind produces an enormous amount of garbage each and every day and that all too much of this was left as visible trash. One had only to take a walk in the woods to find junk automobiles and beer cans everywhere.

The type of life we had been living since the middle of the twentieth century, although very comfortable, also produced mountains of refuse. In the late 1960s and the early 1970s, the feeling that we had become a very dirty bunch had ripped through the hearts of the American people. Suddenly, the Earth seemed awfully small and finite. Our frontiers were gone, and we had to live among the piles of garbage we made every day. We had not only captured and tamed nature, we were doing away with it. Nature was disappearing right before our eyes. The headlines announced that no clean air was left on the face of the Earth; the best air was in the Antarctic, and even it was dirty. Oil spills were occurring two or three times a year. Acid mine water had turned many streams in Pennsylvania into ugly reddish-brown sewers. These streams had once been the home of brook trout, who were good citizens and who could not or would not hurt anybody or anything. We were finding out that mining is a dirty process. Cyanide is bad stuff, and smelter smoke is guaranteed to kill anything. A popular folk song of the time lamented that Mr. Peabody's coal cars had hauled away Muhlenberg County, Kentucky. "And for what purpose?" the environmentalist asked. To haul the coal to Norfolk, Virginia, and to send it across the seas for somebody else to burn. Better to leave Muhlenberg County in Kentucky the way our great-grandfathers found it!

In 1970, one of my colleagues, who was employed by the mining arm of the oil company for which I had worked, went to an environmental impact statement hearing in Tucson, Arizona. He listened in absolute disbelief to the defense used by a mining man who had applied for a mining permit to quarry limestone. It was a misfortune of circumstance that he had proposed to quarry on a hillside clearly visible from a major highway. In the thick of the emotionally charged battle, he said that there was not going to be a problem with his operation being an eyesore because, when he was through, the whole hill would be gone. A nice flat place would be left that would be great for a shopping center. My colleague watched the environmental activists take the quarryman apart. To this day, he probably does not understand that it was not his limestone quarry that was at issue, it was the hillside. What are we to do with men who want to tear down hillsides?

At the height of the awakening concern about our collective untidiness, we had politically active citizens who said that although we might not be able to stop the accumulation of trash and garbage, we must set aside some real estate as wilderness and wild and scenic rivers, expand our national parks, and do whatever we could to hold back the progressive forces of free enterprise from paving over everything with concrete or running over the rest with all-terrain vehicles. We decided that the Earth was fragile and that we had to hem in the construction and mining men and their barbaric machines. We went after the industrial polluters and their solvents that fouled our water supply.

Nothing is ever static. The environmentalists should have known that their strong, successful thrust would soon be blunted by the agents of compromise. After all, some had waged a political battle that was tinged with more than a bit of self-righteousness. The opposition, which came along in due time, made more successful defenses than had the quarryman in Tucson in 1970. James Santini, the congressman from Nevada, would talk to anybody who would listen about why minerals were important and why compromise between environmental concerns and mineral production had to be worked out. My favorite image of him is from an after-dinner speech he gave on a rainy fall evening in 1977 in Washington, D.C., when he said that he could stay and talk all night long about why minerals are important because his basement was flooded and, when he got home, his wife would have a pail and a mop waiting for him. His best line was that a TV set contains 41 components derived from minerals. He said he was tired of trying to be totally rational with the environmentalists on these issues. All he could think of doing was pulling the cord on their TV sets and saying, "No minerals, no TV!" He could have said, "No iron, no automobiles; no magnets, no electricity; or no minerals, no life as we know it!" In 1977, the public was just not quite ready to listen. Most of the economists who listened to Santini's concerns found them curious, if not odd. They could have had fun with him by asking if it were not true that air was the most important element for life, and, if that were so, how come it was free? To the layman, the congressman still had the easily understood point that mineral production, like air, water, and a temperate climate, made for a nice way of life.

Once the counterattack began against the environmentalists, it came from many quarters. Sometimes it came in the form of careful analyses, sometimes as newspaper cartoons, and sometimes as outbursts in the media. One of my most vivid recollections of an outburst was when the president of a major oil company put his foot squarely in his mouth after a major oil spill that his company had caused. He obviously had not consulted his public relations staff before he issued his remarks—he accused the environmentalists of being more concerned about a few oily ducks than about all the people we kill on our highways each day. He actually was more concerned about his company's profits and jobs for its employees than the carnage on our highways. He had forgotten that you just cannot say such a thing to an aroused public. It has about the same result as standing up at a well-oiled cocktail party and yelling out that alcohol and tobacco ruin a lot of lives each year. In both cases, the audience is turned off to the extent that the importance of the long-term problem being addressed by the advocate becomes insignificant.

During the 1970s, the controversy over the withdrawal of federal lands into wilderness areas had developed into a well-defined conflict. On one side of the issue were the environmentalists and, on the other, the promining folks. Both were self-identified combatants, biased beyond repair by self-interest. Who, then, could be objective about the amount of an undiscovered mineral endowment on any given piece of land? There was not much question that because of its policy-neutral character, the USGS had the necessary objectivity to continue making such assessments.

The USGS took on the job of assessing the undiscovered mineral endowment of all the land tracts proposed for wilderness withdrawal as we had taken on similar jobs, such as ensuring a supply of mineral commodities during World War II. During the 1960s and into the 1970s, we worked at the job of wilderness assessments as we thought geologists were supposed to perform their duties. Sometime during the middle 1970s, we started to hear mutterings that our reports were not easy to read or understand. We said, "This is the way geologists write reports. We have done it this way for decades. Our reports are the standard in the profession." Very often, we advocated that further study was necessary before a strong definitive statement could be made about the undiscovered mineral resources of a wilderness area; however, this could not be the whole basis for the rising concern about our products. Quite often, more study *was* necessary, which should have been obvious to anyone who took the time to examine the data.

We had become comfortable with the idea that the geologic data gathered from the examination of each wilderness area ought to be used to state whether there was a high, medium, or low favorability for the occurrence of the various mineral commodities. By using this format specification, we were reasonably certain that we had pushed the data for all they were worth and that the clients would have what they needed in hand. As managers of the Office of Mineral Resources returned from meetings in the DOI, congressional hearings, delegation meetings, and U.S. Forest Service and other federal agency meetings, where the usefulness of our wilderness reports had been discussed, it was common to hear them question each other about the meaning of the frequently cited complaint that our wilderness reports had to be more useful for policy analysis!

As this process of self-examination continued, we concluded that our clients must want to know more than whether or not we had found evidence for the occurrence of mineralization on a tract of land. Almost any piece of land, if large enough, is going to have some mineralization. We knew that if you withdraw enough land, you are going to have to withdraw many mineral occurrences. The probability of withdrawing mineral deposits that could be made into profitable mines also was going to increase as the number of acres withdrawn increased. We could say that lots of things made sense, but what did these people who said wilderness reports had to be useful for policy analysis really want?

The probem we faced was that we could not find anyone in authority who would (or could?) come forth and say, "Now, I do wilderness withdrawal policy analysis, and I tell the Congress how it is going to be done, and this is exactly what I need to decide how to make a land withdrawal decision." Nor could we find any government official who sat in an office that had a sign on the door saying, "Land decisions made here by me on Tuesdays and Thursdays from 10:00 A.M. to 12:00 noon." So, there was no single authority that could tell us what a mineral-resource assessment should have in it or what format the results should have.

We scientists in the Office of Mineral Resources soon realized that the decision-making process for withdrawing federal lands into wilderness areas was not the result of an exact analytical activity. As Tom Ovenshine already knew before he became office chief in 1980, we were finding out that land withdrawal decisions are made in the halls of Congress and in courtrooms. The decision-making process is pounded out by congressmen and lawyers who are after the best deal they can get for their constituents and clients, respectively. It was not science that was being practiced, it was law. We gradually learned that the way to make better, more useful wilderness reports was going to be an evolutionary process. We had produced what we thought was reasonable and were learning from the reactions of our clients, usually indirect and muffled, how to improve our product.

Perhaps the biggest surprise to most of the scientists who worked at producing wilderness reports came on the day they figured out that they were nothing more than expert witnesses in the political process that makes wilderness land withdrawal decisions. We were being treated shabbily in this process because we were expert witnesses, and anyone who has ever been in a courtroom knows that is the way expert witnesses are always treated. What made it even harder for us to understand was that the land withdrawal decisions were made the way civil law suits are settled, where the nearly arbitrary "preponderance of evidence" criterion is used rather than the straightforward "beyond any reasonable doubt" criterion, which is used in criminal cases. Was the near certainty that some sort of mineralization occurs in almost every land tract evidence enough to decide not to withdraw any land tracts into wilderness? On the basis of the type of assessment we could make with the money we were given, we felt that there was still so much uncertainty that it was easy to believe that a significant mineral deposit could be hidden somewhere close to even a minor occurrence of mineralized rock. The rational, scientifically determined verdict could easily come down on the side of saying maybe, just maybe, an undiscovered mineral deposit does occur near the mineralization found in a given field investigation. Before making strong statements, the geologist tends to want more information than we were allowed to acquire in the wilderness program. Drilling core holes would yield the hard test data to determine whether a significant mineral deposit is present or not. This, however, would cost far more than anyone was willing to pay in a program whose purpose was to assess undiscovered mineral endowment; that is, whether a tract of land should be set aside as a wilderness area. The odds, then, were against us from the beginning as we faced the problem of trying to determine what the elusive policy analysts wanted us to put in a wilderness report.

As an illustration of how we learned the manner in which land withdrawal decisions were sometimes made, I offer the experience that a colleague of mine had when evaluating the undiscovered mineral endowment of a proposed wilderness area. My colleague was in the right place at the right time to be given the opportunity to participate in a nearly hands-on withdrawal of a wilderness tract. He had assessed the undiscovered mineral endowment of a rather small land tract that had been proposed for withdrawal. He had concluded from the geologic data available that there was a mineralized porphyry system right in the middle of this tract of land (Ludington, 1984). The exploration geochemical survey gave the telltale positive response, and the presence of intense argillic alteration told him that this tract of land was a hot prospect for drilling by a mineral exploration company. The congressional staffer to whom my colleague presented these results was impressed with this conclusion. The staffer then explained how things work in Washington, D.C. He said that although the analysis showed a high probability for the occurrence of a very large mineral deposit directly under the middle of this land tract, my colleague must understand that although he had provided very important information, the congressman had to weigh many concerns. The staffer pointed out that one of these concerns was that this was the only tract proposed for withdrawal within the congressman's home district. My colleague returned to Reston a wiser man with a broader perspective on how mineral-resource assessments could be applied on public lands.

This is the way it sometimes goes in the big city. From such experiences, we began to recognize that we are just expert witnesses in somebody else's court case. The next step we had to take was to develop enough good sense to shrug off what most of us insular scientists think are poor decisions made for reasons of expedience. My own conclusion was that if we were going to spend part of our professional scientific life being expert witnesses, we must learn not to be offended by what becomes of our contribution in the process of political decision making. In a medical malpractice suit, for example, the physician who testifies as an expert witness for the plaintiff is going to take abuse from the defense attorney. The next day, this physician will be back at his practice tending to sick people. We geologists have to learn to take our turn at writing reports and testifying and behaving the same way.

It is my considered opinion that mineral-resource assessment was on its way to becoming a unified field of scientific activity long before the environmentalists quickened the national conscience against the extraction of minerals from our mines. The basic force that has propelled this field along is the desire to quantify the unknown. We are driven by a powerful urge to look forward. Improving methods for making predictions is, then, a logical consequence of this desire. I try to be positive about the role the geologist plays as a scientist contributing information to such political decisions as the withdrawal of federal lands into wildernesses. For example, when the congressional staffer told my colleague that his data were not going to figure into the withdrawal decision, I rationalized that this was the visible evidence that society had already discounted the value of such a deposit; that is to say, society had calculated an implicit income stream from one more deposit, and its present value was less than the implicit income stream from the alternative use of the land as a wilderness area. Economists say that we make these valuations all the time. To the hard scientist, it is irksome not to be able to find a trace of real evidence that any sort of calculation was made, much less considered. The evidence is there for all to see that decisions are made, but I wonder sometimes if society really gets a chance to consider the meaning of the expected future value of such an entity as the million tons of copper it may have been denied in the withdrawal mentioned above. Of course, some arrangement could be made in the future to withdraw a wilderness back into use for minerals production through a lease sale procedure should the need arise.

At this point, we will turn away from the discussion of the situation in which the USGS found itself as it performed assessments of the undiscovered mineral resources on land tracts proposed for withdrawal into wilderness areas. We now turn toward a discussion of elements of mineral-resource assessment and the development of quantitative assessment techniques. I will start with a statement about where I believe we are today in the application of our science and then make several thrusts back into the past to highlight the three major developments that combined to make up the aggregate body of knowledge and methods. In this discussion, the efforts of the many who have contributed but did not publish their work are unfortunately left out, others are incompletely treated, and some get more than their due by not being criticized. It is my hope that I have had the courtesy to at least mention the majority of the main contributors.

As I see it, the first steps taken in the formation of the field of mineral-resource assessment happened sometime in the unrecorded past when it was noticed that two or more mineral deposits looked alike. The idea then began to form that they might have a common origin. This is, of course, the same beginning that economic geologists claim. In fact, as practiced today, many concepts from economic and regional geology are used in mineral-resource assessment analysis. We achieve our goals by applying these traditional principles that we have extended or grafted onto a statistical framework. By using statistical analysis, we bring to bear ideas and tools that allow us to summarize the character of mineral deposits through common attributes, such as their grade and tonnage distributions. The framework of statistical inference also provides a consistent and robust basis for estimating the undiscovered mineral endowment.

We have progressed along this path to where we base our mineral-resource assessments on the following three-phase procedure: The delineation of permissive regions that uses geologic history, including plate tectonic concepts, intertwined with descriptive mineral deposit models and grade and tonnage models and the estimation of the number of undiscovered mineral deposits, given both of the above constructs (Singer, 1975). Contributions to this three-phase procedure have been made by many researchers who have worked along serial and parallel courses. Although several of these contributions can be documented easily, others take more effort. As mentioned above, progress in understanding why mineral deposits occur where they do has been going on for centuries, whereas the delineation of permissive geologic settings for ore deposit occurrence has been practiced on a large scale only during this century. The first such setting, which was slanted heavily toward the idea that mineral deposits were generated by orthomagmatic processes, was certainly in use in the 1930s (Lindgren, 1933). The use of the principles of plate tectonics to identify geologic elements that explain the regional occurrence of mineral deposits has only occurred since about 1970. For a review of this topic, see Sillitoe (1972), Mitchell and Garson (1981), and Sawkins (1984).

Having been at the scene of the first grade and tonnage model construction in 1971, I can discuss the development of this idea with some vigor (Singer et al., 1975). The development of procedures to estimate the number of undiscovered deposits occurring in permissive terrain has a diverse history and has benefitted from the efforts of many researchers (Nolan, 1950; Harris et al., 1971; Agterberg et al., 1972; Harris, 1973; Singer, 1975; Richter et al., 1975; Page and Johnson, 1977; Singer, 1984; Drew et al., 1986).

The schematic diagram shown in Fig. 10.1 is intended to summarize the paths along which the science of resource assessment has progressed as ideas coalesced to form the three-phase assessment system used today. In this figure, progress in modeling of ore deposits as collections of entities that have similar attributes and the identification of the terrains that are permissive for their occurrence are shown to be closely related. In the practice of making mineral-resource assessments, these two activities are so intertwined that, operationally, they are treated as a single determination. The estimation of the number of undiscovered mineral deposits of a specific type occurring in any region is

Figure 10.1. Schematic diagram showing developments in mineral-resource assessment methods.

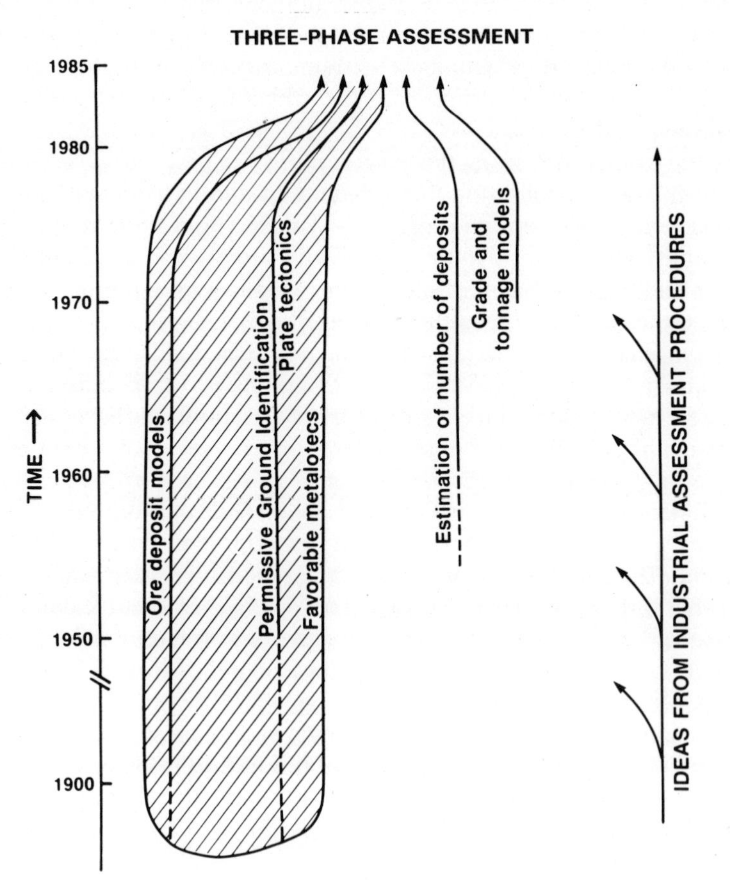

based on weighing the evidence associated with the probability of occurrence of mineral deposits and then transforming these weights to a numerical scale. The grade and tonnage distributions enter the assessment process if and when the probability of occurrence for particular types of ore deposits remaining to be discovered is judged to be nontrivial.

We will examine first the developments made in mineral deposit modeling and the associated activity of identifying the terrains that are permissive for the occurrence of each of the mineral deposit types. At present, we have the state of mineral deposit modeling summed up in atlases, such as those edited by Cox and Singer (1986) and Eckstrand (1984). In both of these volumes, a wide variety of descriptions of mineral deposits are presented. In the atlas prepared by Cox and Singer, statistical distributions for the grade and the tonnage of many types of deposits also are presented.

It is generally recognized that there will always be uncertainties and divergent opinions about which elements ought to be included in a mineral deposit model. These are, after all, models of natural processes. Although we need to be mindful of the philosophical question of what constitutes a model of these processes, it is more important to realize that mineral deposit models are necessary constructs that allow meaningful analytical manipulation of data and efficient summation and communication in mineral-resource assessment.

Paul Barton, a geologist with the USGS, tried to remove the confusion that exists in mineral deposit model terminology by using the diagram shown in Fig. 10.2. This is not a simple diagram because a number of ideas are set forth simultaneously. First, the development of mineral deposit models has been an evolutionary process and will continue to be so. Second, the classification of mineral deposits into groups that have common characteristics is the most fundamental activity in the modeling process.

Once the stage was set with the above two ideas, Barton said that the confusion that often arises in mineral deposit modeling terminology can be removed by identifying the information necessary to obtain different types of mineral deposit models. The most primitive model that can be constructed is a descriptive mineral deposit model, which requires only enough information to identify the common attributes of a collection of mineral deposits (Fig. 10.2). When enough data exist on a large enough collection of mineral deposits that have similar attributes, we may be able to advance to the stage where a grade and tonnage model can be constructed to support the descriptive model. As the information level increases through more detailed investigations and research, we may advance to a genetic understanding of the mineral deposits in the grouping used to construct the descriptive model. At some point, our knowledge may advance to a high enouch level at which predictions of the probability of occurrence of undiscovered mineral deposits can be made from accurate field observations.

Cyprus massive sulfide deposits are a type of mineral deposit often used as an example of a descriptive mineral deposit model because they have been

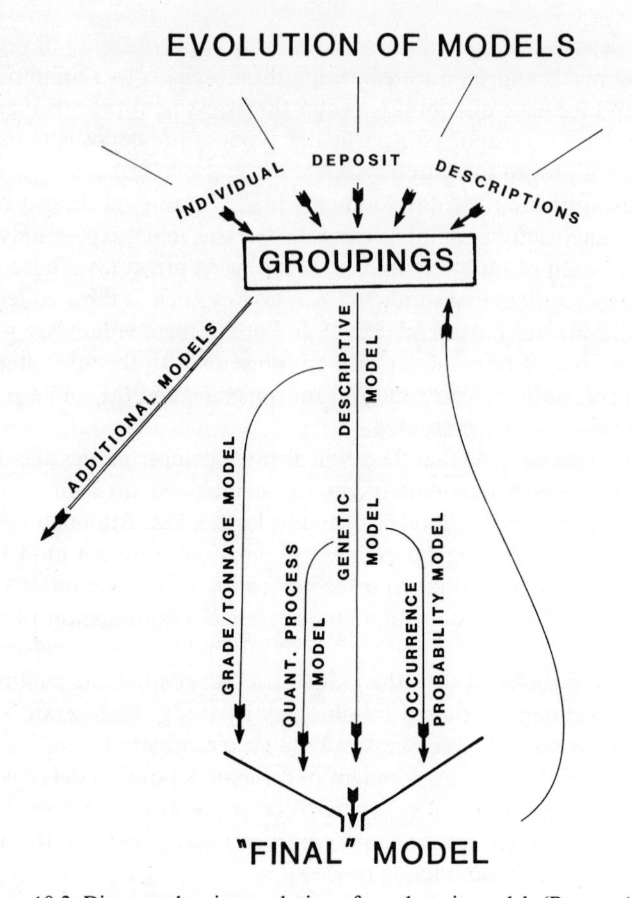

Figure 10.2. Diagram showing evolution of ore deposit models (Barton, 1984).

widely studied (see Table 10.1 for the descriptive model). The critical element in this model is the relation between sulfide mineralization and a stratigraphic position in an ophiolite suite (Fig. 10.3). The geologic history of the type section in Cyprus has been interpreted and summarized in a schematic cross section (Fig. 10.4). In this diagram, the genesis, preservation, and later tectonic movement of the mineralization as a function of plate tectonic criteria are presented. Thus, the descriptive model for a Cyprus-type massive sulfide deposit is a model that has a tightly defined geologic setting (permissive terrain).

When a suspected or confirmed suite of ophiolite rocks occurs in an area, Cyprus-type mineralization and the associated podiform chromite mineralization are candidates for consideration in an assessment of the region's undiscovered mineral endowment. We would say that, by using this level of data, the region is *permissive* for the occurrence of these types of deposits. Evidence of the occurrence of mineralization in the ophiolite suite of rocks obtained from field investigations and/or positive responses from geochemical and geophysical surveys are needed to establish reasonable certainty that the region under consideration is *favorable* for the occurrence of these types of mineral deposits. Favorable means that a nontrivial probability of occurrence has been established.

For a region determined to have the potential for a favorable occurrence of a type of deposit, it may be possible to estimate the number of deposits larger than some minimum size that remains to be discovered. Grade and tonnage distribution data, if available, can be used to transform the assessment of the undiscovered endowment to the more graspable form of frequency distribu-

Table 10.1. Descriptive Model for Cyprus Massive Sulfide

Approximate synonym Cupreous pyrite.
Description Massive pyrite, chalcopyrite, and sphalerite in pillow basalts.
General reference Franklin and others (1981).
Geological environment
Rock types Ophiolite assemblage: tectonized dunite and harzburgite, gabbro, sheeted diabase dikes, pillow basalts, and fine-grained metasedimentary rocks such as chert and phyllite.
Textures Diabase dikes, pillow basalts, and in some cases brecciated basalt.
Age range Archean(?) to Tertiary—majority are Ordovician or Cretaceous.
Depositional environment Submarine hot spring along axial grabens in oceanic or back-arc spreading ridges. Hot springs related to submarine volcanoes producing seamounts.
Tectonic setting(s) Ophiolites. May be adjacent to steep normal faults.
Associated deposit types Mn- and Fe-rich cherts regionally.
Deposit description
Mineralogy Massive: pyrite + chalcopyrite + sphalerite ± marcasite ± pyrrhotite. Stringer (stockwork): pyrite + pyrrhotite, minor chalcopyrite and sphalerite (cobalt, gold, and silver present in minor amounts).
Texture/structure Massive sulfides (>60 percent sulfides) with underlying sulfide stockwork or stringer zone. Sulfides brecciated and recemented. Rarely preserved fossil worm tubes.
Alteration Stringer zone—feldspar destruction, abundant quartz and chalcedony, abundant chlorite, some illite and calcite. Some deposits overlain by ochre (Mn-poor, Fe-rich bedded sediment containing goethite, maghemite, and quartz).
Ore controls Pillow basalt or mafic volcanic breccia, diabase dikes below; ores rarely localized in sediments above pillows. May be local faulting.
Weathering Massive limonite gossans. Gold in stream sediments.
Geochemical signature General loss of Ca and Na and introduction and redistribution of Mn and Fe in the stringer zone.
Examples

Cyprus deposits, CYPS	(Constantinou and Govett, 1973)
Oxec, GUAT	(Petersen and Zantop, 1980)
York Harbour, CNNF	(Duke and Hutchinson, 1974)
Turner-Albright, USOR	(Koski and Derkey, 1981)

Source: Cox and Singer (1986).

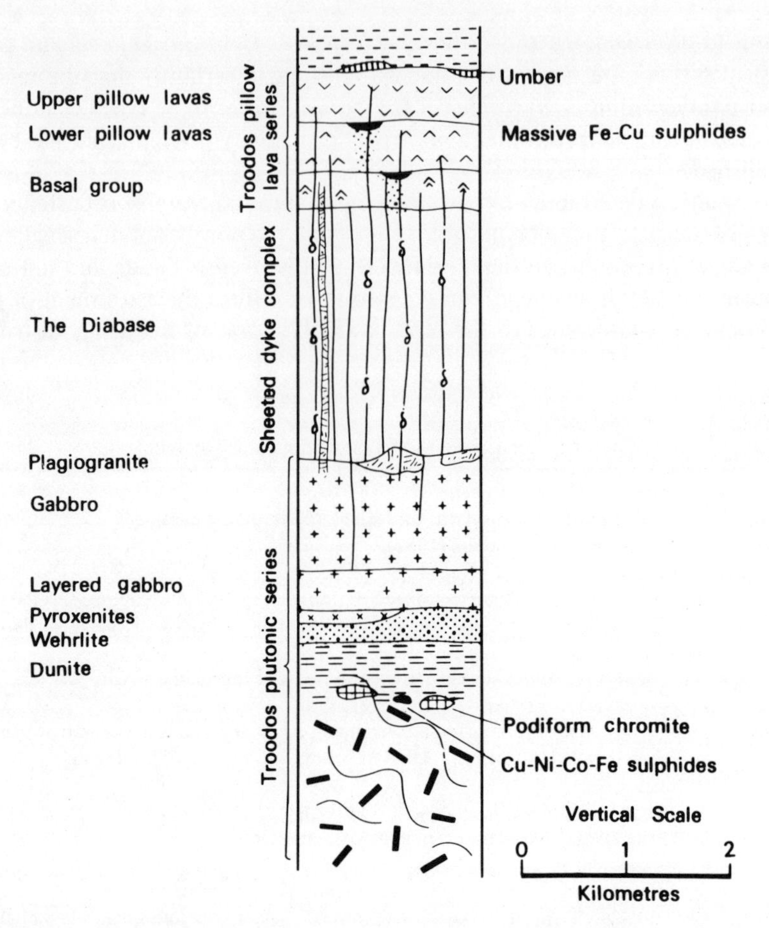

Figure 10.3. Lithostratigraphic section of the Troodos ophiolite (Searle and Panayiotou, 1980).

tions of the tons of metal remaining to be discovered. This is the form that consumers of mineral-resource assessments, particularly nongeologists, seem to prefer.

Many of today's descriptive mineral deposit models are sufficiently well defined and operationally connected to the geologic characteristics of their permissive terrains to be used as a basis for mineral-resource assessments. Perhaps as many as 35 of the descriptive models, along with their associated grade and tonnage models presented by Cox and Singer (1986), are sufficiently well defined with respect to their permissive terrains to be so used.

Developments in this field have been rapid. It has been only a few years since the first attempt to systematically present models of mineral deposits in a standard descriptive form was published (Erickson, 1982). At that time, these descriptive models, although not so labeled, also were closely tied to their permissive terrains. Looking backward to around 1970, we find the first examples of mineral deposit occurrences being tied to plate tectonic concepts. This connection soon led to the conclusion that certain large-scale plate tectonic settings were permissive for certain types of mineral deposits and that compressional plate tectonic settings are permissive for certain types of mineral deposits (such as porphyry copper deposits; Sillitoe, 1972) and are extensional settings for others (such as carbonatite deposits; Sawkins, 1984). If we go back as far as 25 years, we find a revolution in interpretation occurring—the origin of many types of mineral deposits was thought to be hydrothermal replacement instead of being synergenetic. Massive sulfide deposits probably led the way to this revolution. It was the view of some economic geologists at the time (and it continues to be today) that this revolution was, to a significant degree, a revival of older ideas rather than a generation of new ones (Barton, 1986).

Before this revolution occurred, the genesis of mineral deposits was *not* connected so closely to the genesis and geologic character of their host rocks (permissive terrains). I believe that this connection was made at the time because the relation between intrusive rocks and their extrusive equivalents had been established. After this connection was made, the idea that the mineral deposits found in many volcanic rocks were actually a facies of their volcanic host was then quickly established. In the same fashion, the strata-bound mineral deposits that were previously believed to be caused by hydrothermal replacement were reinterpreted as being lithologies in the stratigraphy of volcanic rocks. Ever afterward, we knew that subaqueous volcanic rock stratigraphies were permissive for the occurrence of certain strata-bound mineral deposits that are called "volcanogenetic massive sulfides." To be sure, the really big metal-producing deposits, such as the porphyry copper deposits, were viewed as being orthomagmatic as late as 1960. However, something important had occurred

Figure 10.4. Schematic cross-section to explain uplift of the Troodos ophiolite complex to its present position (Sawkins, 1984).

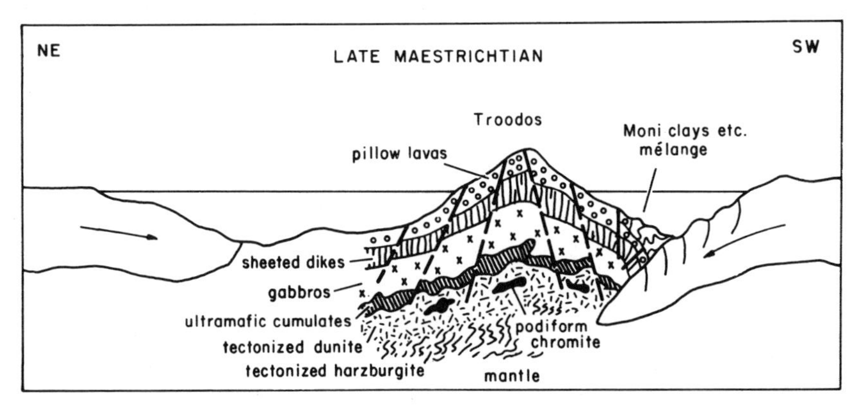

that turned scientists away from the idea that the metal occurring in nearly all mineral deposits was solely the product of the end stages of magmatic differentiation of intrusive igneous rocks. Granites and their kin were believed to play a vital role in the formation of mineral deposits by serving as sources of heat and volatiles, but the idea that much of the metal and the transporting fluids had to come from elsewhere became prominent.

If we go back in time only 55 years to the publication of Lindgen's (1933) epic volume on mineral deposits, we find that syngenetic mineral deposits are limited to magmatic segregations, as in chromite in peridotites, sedimentary iron formation, coal, rock salt, anhydrite, and gypsum. Other than these few types of deposits, there is no mention made of the connection between the mineral deposits and the rocks that host them. This was the era of orthomagmatism and the hydrothermal ore deposit. At this point in the history of economic geology, host rocks were only placed to put igneous intrusions, and almost any type of rock would do as a host rock. Intrusive rocks seemed to occur in mountain chains, and mineral deposits often occurred with them, but not necessarily. Virturally no connection was made among tectonism, sedimentation, magmatism, volcanism, and the occurrence of mineral deposits. Many examples of mineral deposits, such as the tin deposits of Cornwall, England, are found as veins that went out of the granitic intrusive rocks, which were believed to have caused them, into the shales that the granites had intruded into. No mountain peaks were to be seen from the headframes or the pumphouses of these mines in Cornwall, but that did not matter because the intrusive rocks were believed to be the complete cause of the occurrence of these deposits. Consequently, it was not important where intrusive rocks occurred, just so long as they occurred somewhere and that was where the mineral deposits were likely to be found. Today, economic geologists make much of the association between the chemistry and the plate tectonic setting of the basins in which the shale host rocks were deposited and the evolution of the granites and generation of the associated tin deposits occurred. Resource assessment analysts use these associations to define the permissive terrains and to estimate the number of undiscovered deposits in these terrains.

This is an all-too-simple 50-year retrospective glance at the important field of economic geology. It was my intent to point out that we have gone from looking at mineral deposits as being caused, for the most part, by the occurrence of intrusive igneous rocks to looking at the systematic relation between intrusive rocks and the stratigraphy of their extrusive equivalents and derived sediments and at the concepts in plate tectonics that describe the basinal and tectonic settings of the associated larger scale geologic events, which produce a system for the occurrence of many types of mineral deposits. The framework for regional geology provided by plate tectonics has contributed much to this systemization. We can now examine regional geologic patterns and make meaningful statements about certain "packages" of rocks and the types of mineral deposits that they may contain. However, the identification of a permissive setting alone often does not allow us to say with any specific degree of certainty that the geologic processes occurred in the correct order and timing to yield a determinable nontrivial probability that mineralization was created, emplaced, and preserved. Positive site-specific evidence from field investigations is usually required to make this conclusion.

The advances made in the development of descriptive mineral deposit models have also contributed greatly to the systematic development of the field of mineral-resource assessment. These models are much more than a neat and tidy classification tool. They make good scientific sense, and, when tied together with their corresponding grade and tonnage models, they are useful to predict, on the basis of geologic data, not only which types of metals may occur, but also in what volumes. A convolution mechanism, such as a Monte Carlo simulation, can be used to estimate the expected magnitude and the variability of these volumes. Because the application of these models is equivalent to the application of experimental designs, such as those used in agricultural yield experiments, the descriptive models are then comparable to the treatment effects in such experiments. It must be remembered that, in this equivalence, the mean and the variances vary across the treatments.

We now move from a discussion of descriptive mineral deposit models to one of grade and tonnage models and then to some ideas on how to estimate the number of undiscovered deposits that occur in a region. Some of the history of their development and early application is also presented. Grade and tonnage models are discussed first because they are a prerequisite for the procedure used to estimate the number of undiscovered deposits. In fact, the concept of a grade and tonnage model is deeply embedded in the procedure. Grade and tonnage models are not, as is often thought, a device added onto the end of a mineral-resource assessment only for display purposes as a sort of afterthought.

Sixty grade and tonnage models, along with their descriptive mineral deposit models, are described by Cox and Singer (1986). The simplest of these models describes situations in which a deposit type contains only a single mineral commodity and the grade of the ore is independent of the ore tonnage across the deposits. When grade and tonnage models were first constructed, this was a common representation; for example, the first model for porphyry copper deposits contained only information of the distribution of copper grade and ore tonnage because data byproduct metal grades were not available (Fig. 10.5). One of the important conclusions from the analysis of the structure of this model was the independence of the ore grade versus the ore tonnage. This conclusion was interpreted to mean that the future supply possibilities for copper from this type of deposit did not include the continuous discovery of deposits with lower and lower grades. If this were true, as had been assumed, the future supply of copper was only a matter of increasing the economy of scale of the mining operation.

Today, the grade and tonnage model for this deposit type is more complete and is displayed in marginal distribution format (Figs. 10.6–10.8). In this rep-

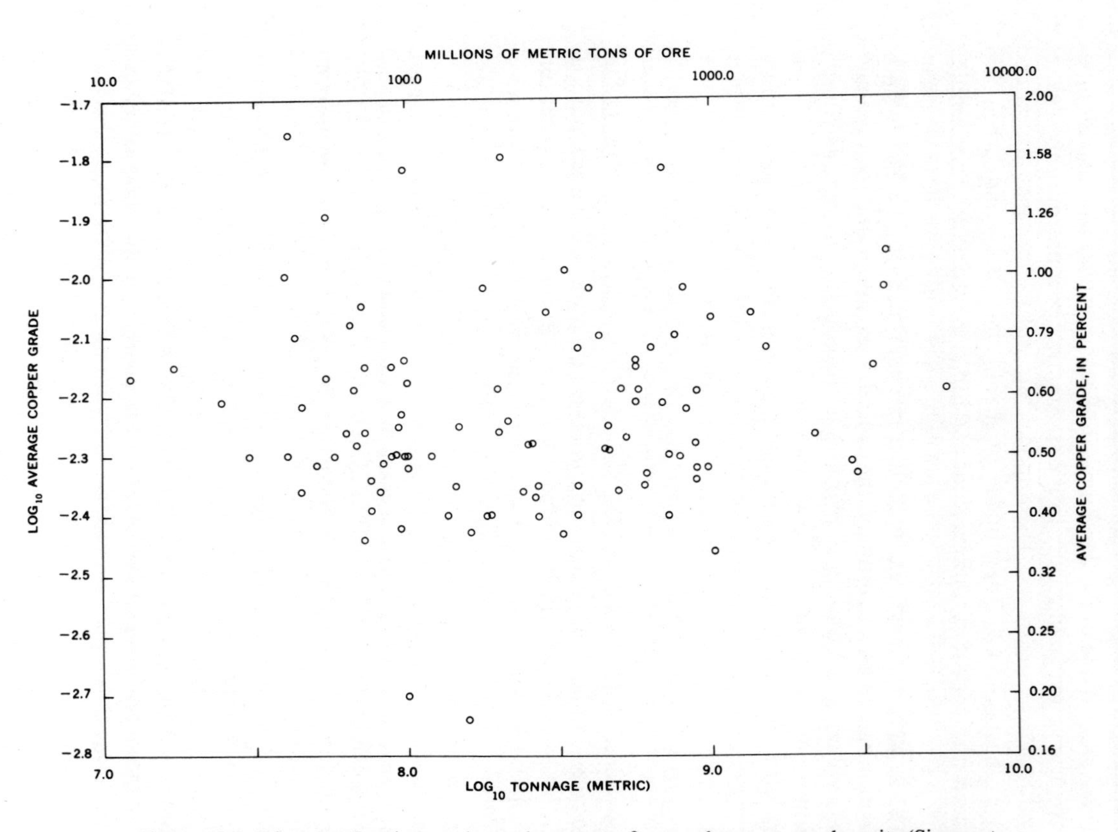

Figure 10.5. Diagram showing grades and tonnages for porphyry copper deposits (Singer et al., 1975).

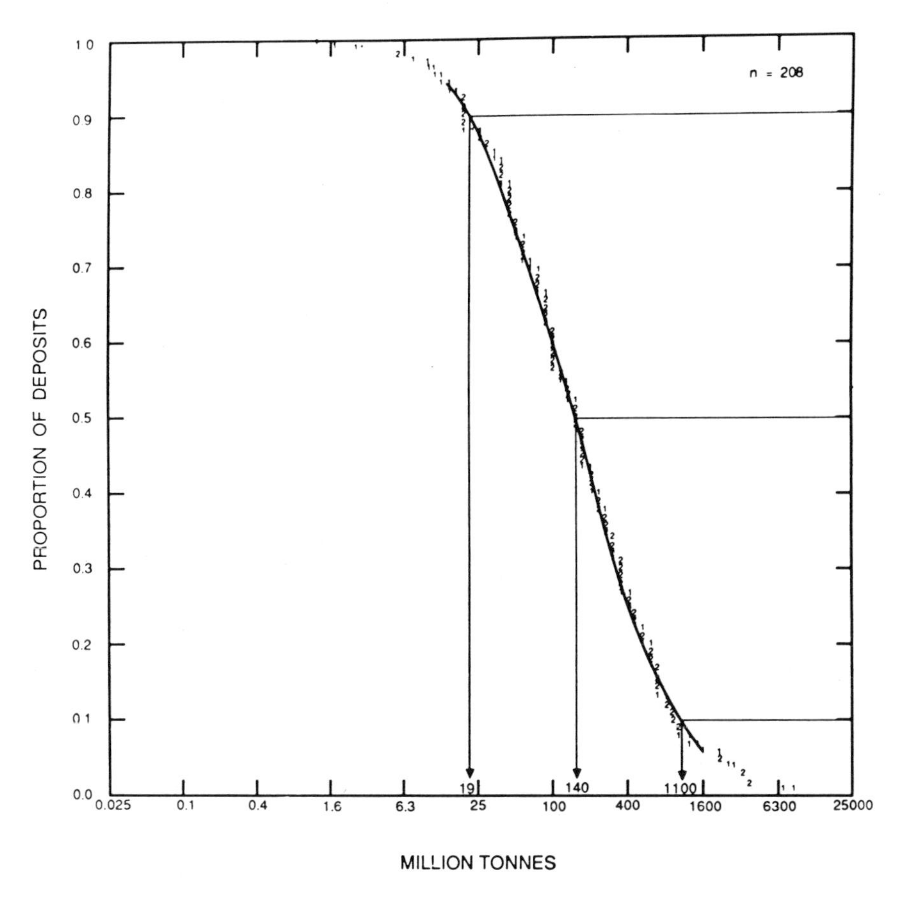

Figure 10.6. Cumulative distribution for tonnage of ore in porphyry copper deposits (from Singer, Mosier, and Cox in Cox and Singer, 1986).

resentation, the marginal distributions for the byproduct grades of molybdenum, gold, and silver are included in addition to the distributions for the copper grade and the ore tonnage. The correlation structure for each of these variables is also available, as are the marginal probabilities for the existence of these byproducts (Singer, Mosier, and Cox, in Cox and Singer, 1986); for example, the probability that recoverable gold and silver are present in a porphyry deposit is about one-third. From the correlation structure, we see that the gold grade is inversely correlated with ore tonnage ($\hat{r} = -0.49$, significant at 0.05 level) and with the molybdenum grade ($\hat{r} = -0.45$, significant at 0.05 level). The copper grade and ore tonnage remained uncorrelated as new and more complete data for this deposit type were collected over the ensuing decade (1975–1985). The progressive movement toward more complete and usually more complicated grade and tonnage models is, for the most part, a natural

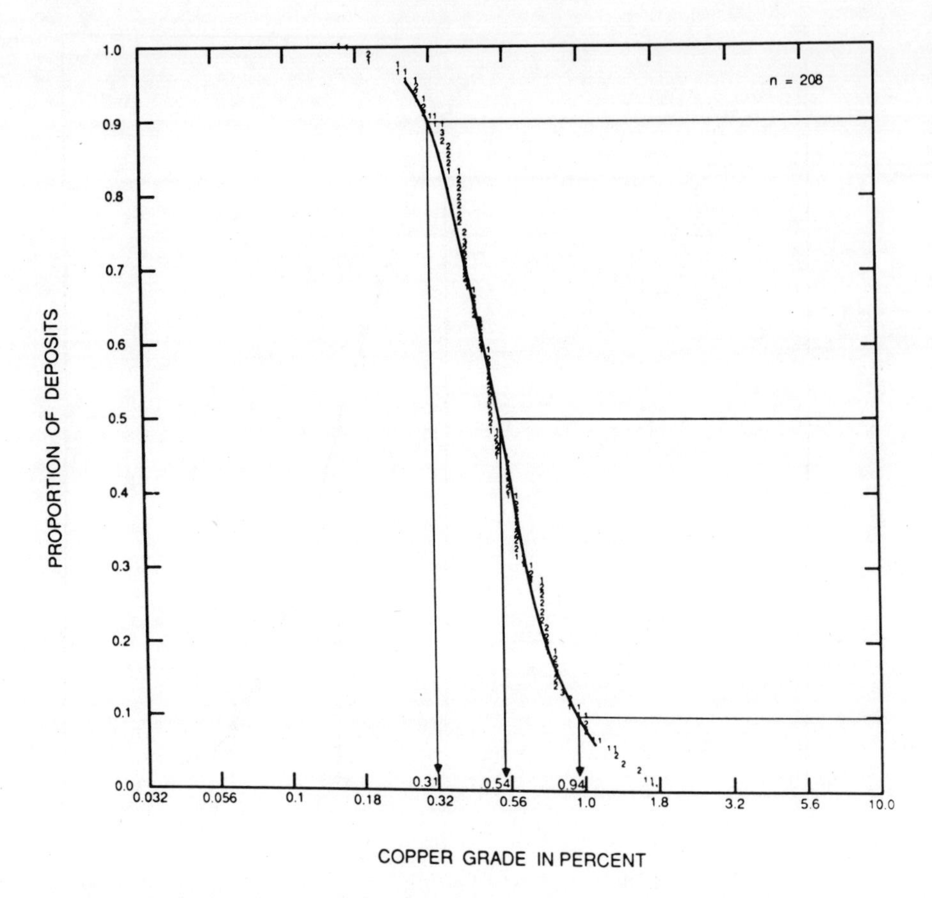

Figure 10.7. Cumulative distribution for copper grade in porphyry copper deposits (from Singer, Mosier, and Cox in Cox and Singer, 1986).

consequence of the process of gathering more data at the same time that we are learning more about the internal structure and geologic settings of mineral deposits.

When we briefly examined the origins and development of descriptive mineral deposit models above, we found a somewhat diffuse history going backward over many centuries. Much of what happened has had to be pieced together. Because we do not have the personal journals of the participants, it is difficult to chronicle the manner in which events unfolded in this field. However, the origins and the time path of development of grade and tonnage models are so recent that nothing more than stirring the memories of the living needs to be done to discover how the field originated and why it developed as it did.

Having participated in the beginnings of these models, I thought that it

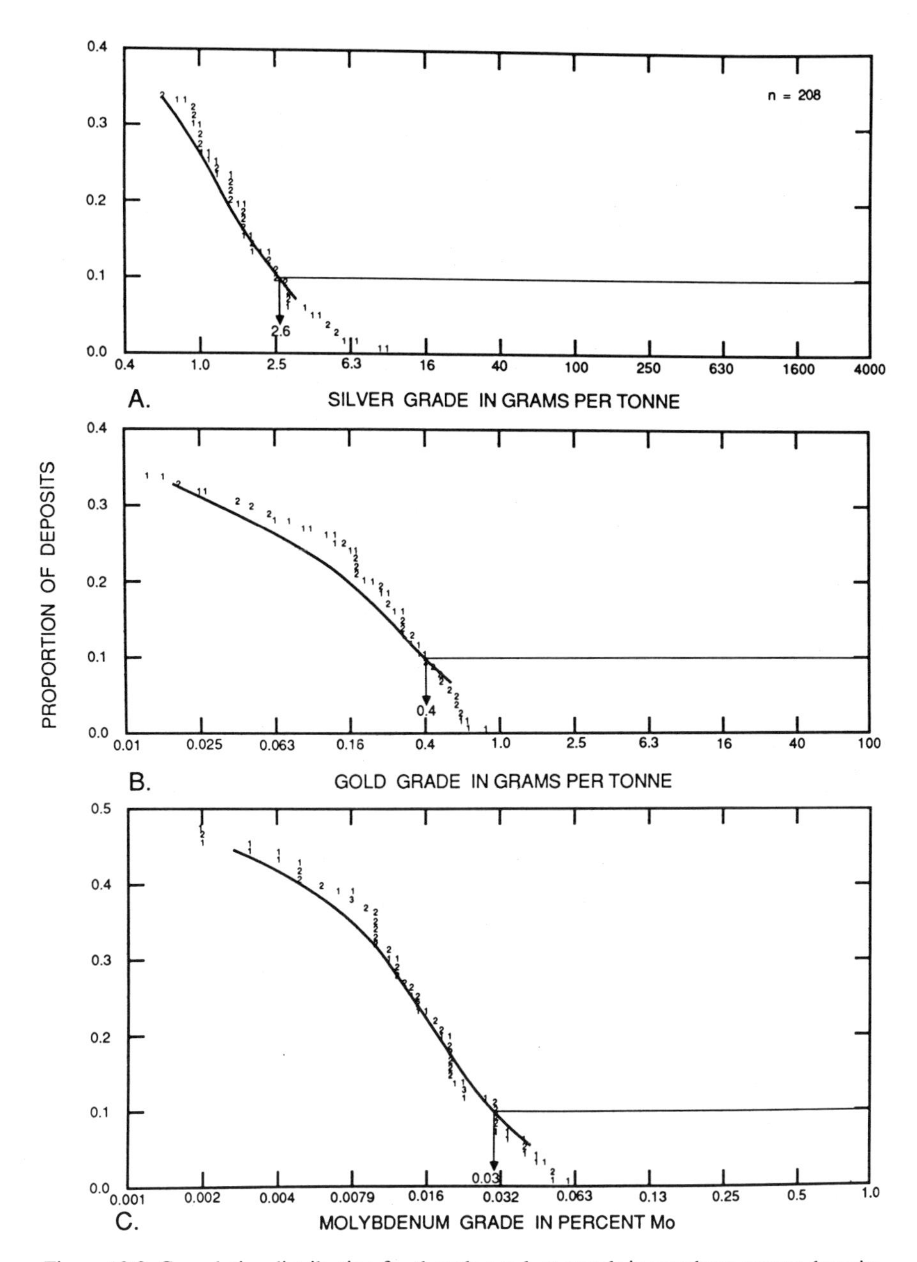

Figure 10.8. Cumulative distribution for three byproduct metals in porphyry copper deposits (from Singer, Mosier, and Cox in Cox and Singer, 1986).

would be easy for me to write down the reason for the early work on them. However, it took more effort than I thought it would because one must try to remember accurately extremely detailed facts. The tale begins sometime in late fall 1971, when Don Singer and I independently took pieces of graph paper and plotted estimates of the size (tonnage) against the average grade of the porphyry copper deposits in the United States and Canada. We plotted them on a logarithm-by-logarithm scale. In my case, necessity was the mother of invention. I needed a device for a Monte Carlo modeling project requested by the mining arm of the oil company for which I worked at the time. This device had to yield a point estimate of the grade and tonnage of the deposits that had been discovered when all the probabilities in the simulated chain of exploration activitics had led to success. I needed a device that showed what success in mineral exploration looked like.

After a meeting with company officials in Salt Lake City, Utah, where I had gone to report my progress on the simulation model, I had occasion to contact Don Singer, who was then working for a mining company in that city. The purpose of my call was to look up a fellow graduate student whom I had not seen for awhile. To this day, I remember the look of surprise on Singer's face after I opened my briefcase to answer his question about what I was working on and why it had brought me to Salt Lake City. I pulled out my pieces of graph paper and said that the grade and tonnage statistics for copper porphyry deposits had not turned out as I had expected. Looking down at the scatterplots on which the deposits owned by my company were plotted in green, the other deposits in the United States plotted in blue, and the Canadian deposits plotted in red, Singer remarked that, when I computed the correlation coefficient, I would find it to be nonsignificant.

Now how did he know that? I wondered. The way he spoke told me that he had not instantaneously deduced this conclusion upon seeing my scatterplots. He told me that he had recently made the same scatterplots. Although he was almost certainly violating his company's security policy by making such remarks, he was most careful to avoid saying why he had made these same scatterplots. It was a different matter for me because I was a quasi-consultant of the oil company's research laboratory. I had been given a couple of months to throw together a computer model that described the mineral exploration process. When this computer model had done its job, which was to teach a little respect to some money managers in the corporate headquarters in New York City, I would be going back to the oil and gas business. Don Singer had to obey the rules of the culture in which he lived. He was much closer to where real dollars were changing hands than I was. When you get close to real dollars in the exploration business, lips are buttoned. So I never knew why he plotted his points on his graph paper, and, to this day, I have never asked. After all, I never needed to know. What is curious is that the two of us to the year, maybe to the month and maybe to the day, had the same idea. The data we had plotted in

1971 were published later, under coauthorship with Dennis Cox. With minor alterations, it is displayed in Fig. 10.5.

Working together during the middle 1970s at the USGS, Don and I would find grade and tonnage models useful in what was, for a time, one of the major controversies over the adequacy of domestic mineral supply. We were embroiled in an argument that seems rather silly today. On one side, there was a group of energetic folks who argued that this nation could never have a mineral supply problem. To them, history had demonstrated that to have an adequate supply was only a matter of lowering the grade of the ore mined. As the average grades fell, new supplies would continue to pour forth; that was all there was to the matter. It was an open-and-shut case closed! On the other side was a not-so-noisy group composed mostly of earth scientists, including Don Singer and me. We questioned the notion that you could at any time lower the grade of the ore mined, thus causing the mines currently in production, along with the new ones that would be discovered, to produce more and more copper, lead, zinc, and so forth. We saw a powerful article of faith being applied by those who argued that the cornucopian viewpoint based on a function of lowering the average grade was valid.

As many other geologists had done before, we pointed out that the biggest copper deposits were the porphyry deposits and that they were currently the principal type of deposit being mined. From our grade and tonnage models for these deposits, we argued that the average grade of the ore and the recoverable tonnages across deposits were independent, which meant that dropping the average grade to achieve more mineral production was not really under our control. We could certainly mine lower grade material in the United States, but there was not an enormous volume of copper resource ready for the taking just by turning the dial downward toward a slightly lower grade in the mines that were in production or by lowering the cutoff grade for the development of newly discovered deposits. We asserted that the independence between grade and tonnage in the porphyry deposits implied that, if this argument were true, a new type of copper mineral deposit would have to be found that had a lower grade and higher tonnage than the porphyry deposits. We could not see how that could occur. We asked what type of deposit it would be and under what geologic conditions it could have formed. We also were ready to question the idea for other mineral commodities.

However, as with so many of the debates in the 1970s involving issues in the supply of fuel minerals and nonfuel raw materials, market forces were at work that, at least for some time to come, would render this entire subject a nonissue. By the middle of the 1980s it had come to pass that the world had more base and ferrous metals in its markets than it knew what to do with.

This abundance can be traced to supply and demand. On the supply side, the availability of metals, such as copper, zinc, tin, and iron, expanded as production increased, which was the result of new discoveries and enhanced production from existing deposits around the world. On the demand side, per capita consumption of metals fell. One obvious reason for the decline in consumption was that in the 1970s gasoline prices rose, which forced the weight of automobiles to decline. Smaller automobiles meant a decline in the use of many types of metals. During this same time, other types of physical and technical substitutions for metals were made as a result of concerns for the environment (for example, lead in gasoline) and changes in taste.

Tin became an often-mentioned example of a metal whose time of prime use had passed. Another metal, aluminum, and plastics had been widely substituted for tin. In addition to this decline, the situation in the tin market was further disrupted by the artificially high price put in place and maintained by the monopolistic activities of a producing cartel. The noncartel producers of tin expanded their production as the cartel fought to maintain the price by stockpiling. A crash in the price had to occur. The cost of the stockpile, which in the middle 1980s had grown to about 100,000 metric tons from a normal trading stock of about 20,000 metric tons, could no longer be financed by the cartel. The end of the cartel's hold over the price of this metal came in the fourth week of October 1985, when the price fell by two-thirds. At that point, the fate of the tin producers was the same as that of the producers of other base metals—low prices would place pressure on producers to lower production. By the middle 1980s, the double-edged sword of expanded production and lowered consumption of most metals had taken away the concern about adequacy of supply of these raw materials that had existed in the early 1970s.

It is my firm belief that mineral production capacity of the world increased on the supply side as much as it did in the 1970s and early 1980s because of the pure economic motives of the developing countries that worked in concert with the political and economic motives of the developed countries. There is a symbiotic relation between these two groups over the issue of minerals production. The developing countries used mineral exports for foreign exchange, and the developed countries wanted to deny the suppliers of mineral commodities the possibilities of maintaining or gaining additional power over the price of these commodities. In this relation, technological know-how and economic aid flow from the group that had an interest in breaking monopoly power (developed countries) to the group that sought foreign exchange through development of what was, in the 1970s, their substantial endowment of conventional deposits of mineral resources (developing countries). For the developed countries to foster development of mineral-production capacity around the world is an effective strategy to deny monopoly power over the price of metals. The effects of the 1973 oil embargo on the economies of the developed countries were, if possible, not going to be repeated in the metals market!

Since the early 1970s, the result of this symbiotic activity has, for the most part, turned the world's mining industry into a competitive industry; that is, the mining industry today is like the troubled American family farm, which is laboring in a competitive market and is burdened with overproduction and inelastic aggregate demand. In this situation, normal profits are not present, and producers are under pressure to shut down. To the buyers of the output from such sectors of the economy, this is the best of all possible worlds—raw materials are cheap and abundant.

Our new tool, the grade and tonnage model, had been denied an opportunity to contribute to the resolution of the debate over whether the average grade of the ores mined would slowly decline, thereby opening an ever-expanding supply of metals. The oversupply of most metals on the world metal markets had reduced this burning question of the early 1970s to a nonquestion by the middle 1980s. We did not despair for our new tool because in the meantime the new field of quantatitive mineral-resource assessment would provide an application for grade and tonnage models far beyond the resolution of the debate over whether the supply of mineral ores could be ensured by mining what amounted to an infinite supply of ore by doing nothing more than extracting it as the grade fell gently. Instead, we were going to contribute to the problem of placing a value on tracts of public land that would figure into decisions about how that land ought to be used in the future.

Under the leadership of Don Singer, many of us at the USGS began to realize that we should build our quantitative assessment methodology around the double-barreled concept of descriptive mineral deposit models and their associated grade and tonnage models. It was quickly recognized that the grade and tonnage model provided a quantitative scale that allowed a metric to be placed into the assessment methodology. In other words, grade and tonnage models provided the standard against which the magnitude of favorability for deposit occurrence could be measured. In the parlance of the assessment analyst, when we report a "nonzero" expectation for the occurrence of a type of mineral deposit, we are saying that we have sufficient evidence to be *on* the grade and tonnage model associated with that mineral deposit.

A schematic diagram of a grade and tonnage model is displayed in Fig. 10.9. In this figure, two domains, A and B, are shown. Their combination defines the total domain for the permissiveness of occurrence of a mineral deposit type attributable to a region. Only in domain B are the combinations of grade and tonnage high enough to specify a deposit that is similar by our definition to those currently being considered for mining. When a mineral-resource assessment is made, the number of undiscovered deposits expected to occur and the combinations of grade and tonnage in domain B are of central concern to an assessor. The reporting of an expectation that no undiscovered deposits occur in a permissive terrain means that all the undiscovered deposits expected to occur have grades and tonnages in domain A.

When I became branch chief in March 1982, I knew that I was expected to move the field of mineral-resource analysis toward further quantification and, in particular, toward the goal of estimating the tonnages of contained metal remaining to be discovered on tracts of federal land. However, I had no inkling that I personally was going to become involved in the assessment of the undis-

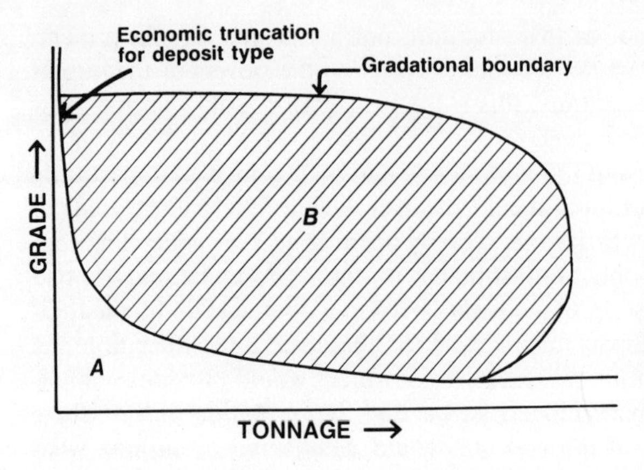

Figure 10.9. Schematic diagram of the domains of a grade and tonnage model: B is the domain of the model, A is the domain where mineralization occurs but has never been mined.

covered mineral resources in the wilderness areas of the Pacific Mountain System and the estimation of the undiscovered tonnages of metal by estimating the number of undiscovered deposits. I found this out at an office management meeting in September 1983. It was true that, years before, I had worked on the construction of the first grade and tonnage model and had enthusiastically followed the development of descriptive mineral deposit models by Don Singer, Ralph Erickson, Roger Eckstrand, Dennis Cox, and many others. All I knew was that it was time for our yearly management meeting, which had been convened in Marquette, Michigan. The meeting had been in progress hardly an hour before I had an assignment to produce a numerical forecast for the tonnages of metals occurring in the deposits yet to be discovered in a collection of land tracts. Once again, I relied on the Monte Carlo simulation to produce the required summation.

As I recollect, it was not only cold in Marquette on the morning of September 20, 1983, but it was snowing as Tom Ovenshine moved the meeting through the initial agenda items. It had become routine in these meetings to be treated to a lecture on being prompt in the completion of mineral-resource assessments on our wilderness area and two-degree sheet assignments. The point was repeatedly made that deadlines were to be respected, even honored. In the midst of making this familiar admonition, Ovenshine suddenly took a detour to a recapitulation of the history of mineral-resource assessment as we had practiced it over the last two decades.

I had done what one normally does at these sorts of meetings when scoldings are being delivered. I had sought out and found that semiconscious state of mind so useful for enduring most, if not all, management meetings. The absence of a window from which to glance from time to time and the stuffiness in the room had ensured an even lower-than-normal state of mental activity. As Ovenshine held forth, he touched on the issues with which he had been faced in bygone years, such as reporting our assessments of undiscovered mineral endowments in a "high-medium-low" format. He discussed the more site-specific assessments in which land tracts were identified as being permissive and/or favorable for the occurrence of particular types of mineral deposits. The maps that displayed this type of assessment had been prepared initially for the Alaskan Mineral Resource Program (AMRAP); because of the rather weiner-like shape of many of the permissive and favorable tracts, these maps had come to be called "AMRAP hotdog maps." Tom liked to use this phrase. He also pointed with favor to descriptive mineral-deposit models and their grade and tonnage model equivalents. The recent yet-to-be-published contributions of particular individuals were mentioned (Hodges et al, 1984), and our substantial progress to date was praised. All good stuff. Tom was very good about finding the positives in nearly any issue. As he talked on, it seemed that we were being treated to a nice pep talk.

He then suddenly turned up the volume and announced that there was more to be done! This was not a good sign, and a small stir went through the room. There is always the worry that behind the façade of a pep talk lurks an assignment. Sure enough, he looked at me and said, "Now, Larry Drew is going to tell us all how we are going to take our next big step toward our goal of quantifying mineral-resource assessments." I think I said with modest emphasis, "Who me?!" The equilibrium of the meeting had been disturbed.

Ovenshine wanted further progress in the technology of areal resource assessment and had used the tactic of the surprise attack to start the ball rolling. Moreover, he knew exactly what he wanted, and, now that he had our attention, we were told his desire at point-blank range. He wanted favorable terrain analyses combined with grade and tonnage models, and he wanted the number of undiscovered mineral deposits explicitly estimated. He also wanted these ideas to be melded together to produce a type of assessment that was graspable by the audience of concerned people who occupied positions in all facets of government, in addition to those in the private sector. Two years later, he wrote down his ideas for all to read (Ovenshine, 1986). At that moment in Marquette, we branch chiefs started to scramble for position. Rapid backfilling was the immediate response with which Tom had to deal. A common voice was heard saying that this was not reasonable because the problems with data alone put this goal of estimating the number of undiscovered deposits beyond solution. Tom knew that was going to be the consensus response, and all he wanted from me was a commitment that I would consider his request seriously enough to do something about it. The shock treatment was for my counterpart branch chiefs. Its purpose was to create a little breathing space. Tom knew that they would soon swarm like bees and try to disable his idea.

Tom Ovenshine believed that, sooner or later, mineral-resource assessments were going to have to be done by using estimates of the number of undiscovered deposits so that the economics of exploring for mineral deposits in the future could be estimated and used in making decisions about the general welfare. His intention was to get out in front on this issue so that his Office would be in a lead position. He called his proposal the "Mark III Project." This label survives today as the name of the Monte Carlo simulation program we used to estimate the tonnages of metal contained in the undiscovered deposits occurring in our assessment of the mineral resources of many different types of land tracts.

Back at our branch headquarters in Reston, I told the more adventuresome geologists, mathematicians, and economists about our new-found objective. They concluded that I should go back and quietly ask Tom what compromise he might be willing to accept. When I did, he took the recently published 1100-page Professional Paper 1300 entitled *Wilderness Mineral Potential—Assessment of Mineral-Resource Potential in U.S. Forest Service Lands Studied, 1964–1984* (Marsh et al., 1984) from his shelf and said that it was a nice book for one geologist to discuss with another, but it was not a vehicle to convey the meaning of mineral resource assessments to the nongeologists of the world. That, however, is what it was supposed to have been. The point was made again that we must produce our assessments in the future and transduce what we had done in the past into a graspable form. There was that word again: "graspable!"

Tom's point was that the "if" and the "maybe" and the "where" of production from mineral deposits that had not yet been discovered was our business. Also, it was not foolish for an economist or policy analyst or any concerned person to ask such a question as, "How much cobalt will be locked up in the U.S. Forest Service wilderness areas?" To Tom, the question was to be answered in the sense that the question was asked. A simple, meaningful statement was required, not a long discourse on the geology of the rocks from which cobalt might be produced someday in the United States.

His point was clear enough to me. We were going to do something about producing our mineral-resource assessments in a form somebody else could use to evaluate the consequence of a land-use and/or withdrawal decision. To make this contribution, we had to produce mineral assessment information in a form that could be transformed into an economic statement. Estimation of the expected number of the undiscovered mineral deposits occurring in a region was the obvious first step in this direction. Data in this form had to be produced so that the trade-offs between the use of a piece of land to generate value to society from future mineral production and the values a piece of land could generate from its alternative uses could be computed by a decision maker. This was an easy enough problem to state but one not so easy to solve.

After a month or so, Tom asked how his Mark III Project was coming along. My answer was that I was thinking about the problem of whether, on average, a piece of ground that contained two square miles of permissive or favorable terrain contained twice as many deposits as a similar one-square-mile plot of ground would. I drew graphs of linear and nonlinear functions and pointed out that it was not obvious how to proceed. I suggested that a spatial mineral deposit occurrence might, in general, best be characterized as contagious. Could it be that if you find one deposit, the probability is high of finding another deposit close by, and when no mineral deposits have been discovered, the probability is low that a discovery will be made? I knew this was a weak response and was only going to hold him at bay long enough to make an exit. The problem that had been placed before me at Marquette had quickly turned itself into a conundrum. I could see only a tangle of opinion before me as we discussed how to estimate the number of deposits remaining to be discovered in a land tract. I decided to stall for time.

Within another month, Tom's question changed from "How are you coming on my project?" to "Where is my report?" It had not taken any great genius to see that the end of the rope was in sight. I returned to the project geologists and brought the news that it was time to assemble, to select a study area, and to make a numerical forecast of the undiscovered endowment by deposit type. Before we were through, the team would grow to consist of a dozen members; it had started modestly enough with four economic geologists (Dennis Cox, Norman Page, David Menzie, and Donald Singer) and myself.

The test area chosen was the collection of U.S. Forest Service wilderness areas in the Pacific Mountain system (Fig. 10.10). This collection of areas is located along the entire length of the system and includes diverse geology. Whether the choice of these areas was fortuitous or not, we can look back and wonder if it was not the result of a carefully laid out plan. If not, then it just so happened that the four economic geologists could take various subsets of these wilderness areas and be prepared for the elicitation phase of the assessment process within three weeks. In this phase, the numbers of undiscovered mineral deposits of each type expected to occur were estimated. In the northern part of the Pacific Mountain system, a subset of wilderness areas (in the State of Washington) was determined to be favorable for the occurrence of porphyry copper deposits. Dennis Cox, the copper commodity specialist, took the lead for these areas. Located to the south of these areas is another group of areas (in Oregon and northern California) that were determined to be permissive for the occurrence of podiform chromite, massive sulfides, and small manganese deposits. Norman Page, who had substantial field experience in these areas, led the assessment procedure for these regions. David Menzie, who had experience with precious-metal mineral deposits, led the effort for those wilderness areas that were located in the mother lode country of California. Donald Singer, the fourth economic geologist on the original team, took the collection of wilderness areas in central and southern California where positive evidence for the occurrence of mineralization was sparce. My role was to serve as manager and to write the Monte Carlo simulation used to calculate the metal endowments and the final report.

In principle, the group did not object to estimating the number of undiscovered deposits occurring by deposit type. The collective worry was that we were going off across the poorly charted interdisciplinary field that is located between economic geology and applied economic analysis; there are very few landmarks to be found on this journey. We were also headed in the direction of nonfuel

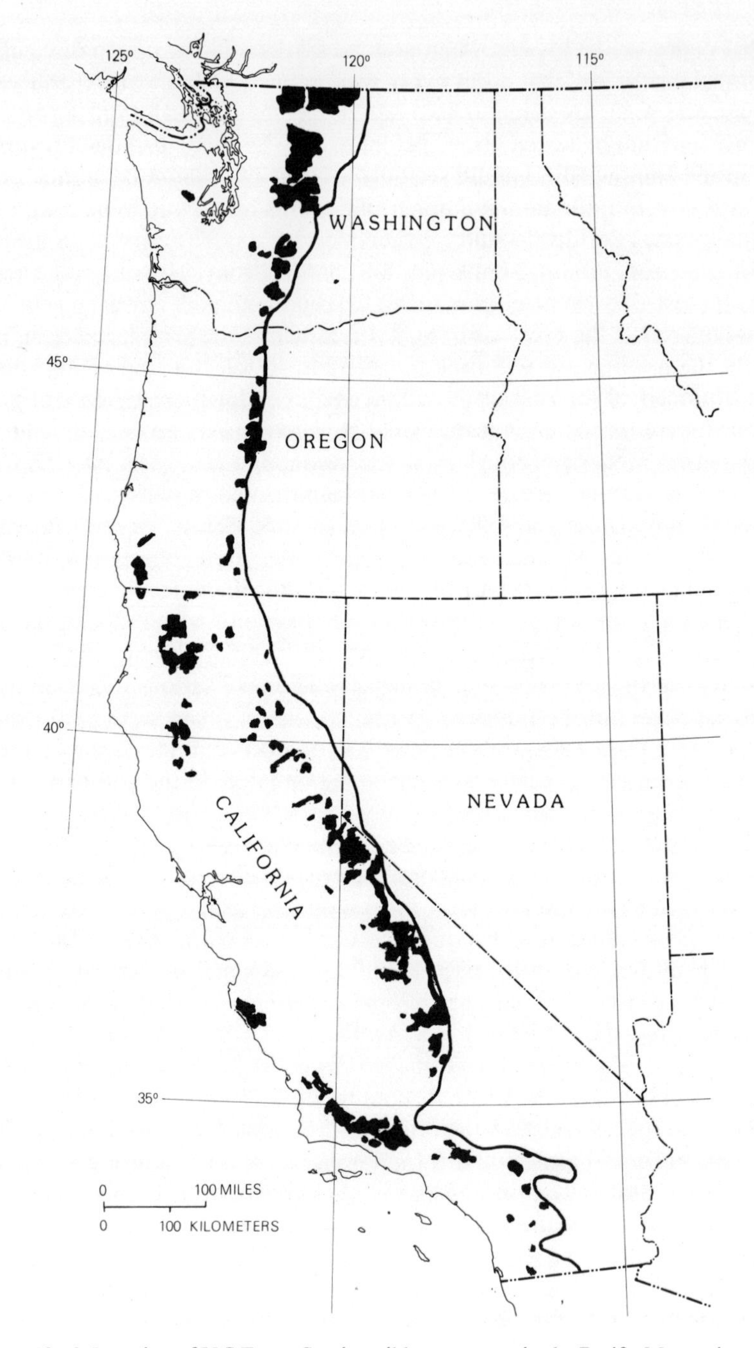

Figure 10.10. Location of U.S Forest Service wilderness tracts in the Pacific Mountain system.

mineral policy analysis, an environment that, when penetrated, has no more landmarks for the geologic mind to use than were encountered on the journey to it.

As we worked our way through the interactive phase of the assessment, we were confronted with situations and conflicts that, when resolved, tended to raise our confidence in the outcome of the assessment. In general, our confidence rose as the group members realized that the projection of logical statements, ideas, beliefs, and feelings onto a numerical domain opened the way to the most efficient form of summation and communication. To those scientists who practice their trade by using the language of mathematics, making this statement is nothing more than restating the obvious. To the pure mathematician, every verbal statement has its direct mathematical equivalent. If it does not, the mathematician asserts, the statement is nonsense. The notion that verbal statements have numerical equivalents is odd to any scientist who believes that concepts and data are to be manipulated mostly as collections of words peculiar to the dogma and the jargon of their professions. Through mindset, training, or preference, geologists usually avoid the language of mathematics and, therefore, fall into this group.

Holding such a position generates the propensity to avoid making predictions in the domain of realtime. To me, it is odd to observe the contrast between the total comfort that geologists feel when they discuss such topics as the changes in mineral phases in rocks as a function of temperature and pressure across the vagueness of the geologic past and the lack of comfort that they feel when they take the risk of being caught making realtime predictions about geologic phenomena, the outcome of which might be subject to a test in the immediate future. The trick is to be able to move to a position where historical proof is not required, and the vagueness of the geologic past is then nullified. I think this conclusion not only holds true for the prediction of the size and the character of the undiscovered mineral endowments that could be subjected to the test of the exploratory drilling program of tomorrow, but also for a much wider spectrum that includes the prediction of the occurrence of earthquakes and the 100-year flood.

When a forecaster such as myself pushes a prediction into such an environment, as we do with our mineral-resource assessments, I am subject to the criticism that, to my mind, is largely the result of being too specific about what may occur in the future. It might be acceptable if the assessment of the undiscovered mineral-resource endowment of a region in terms of the number of deposits expected to occur by deposit type would be the final statement. The problem comes with the realization that it is only a short hop, skip, and jump to the question of how long it will take to find the undiscovered deposits that are expected to occur. In general, our answer would be that we cannot really say; however, we would expect that discoveries could be made if a certain amount of money was invested and if a certain number of exploratory drill holes were drilled. Even though we have not marked an exact date on the calendar, such statements indicate that we have gone too far toward associating the endowment estimation problem with the dimensions of realtime. The fear of using the time domain may be explained as distaste for the idea that the resource assessment problem is a forecasting problem that can only be handled in a meaningful way by use of a statistical analytical structure.

The battle lines usually are drawn up over the issue of availability of information that is needed to make respectable forecasts of the undiscovered mineral-resource endowment of any region. This is a weapon that can always be called on because no matter how much effort is put into the study of attributes of mineral deposits and their spatial occurrence, the judgment can always be made that more data are needed before it is safe to make a prediction. I see this device being put forth to confuse and intimidate forecasters by trying to increase internal doubt to the point where we will cease and desist of our own accord. The order to stop making estimates of the number of undiscovered mineral deposits is, of course, never given. The tactic seems to be to stifle the statistical realism in the issues of mineral deposit modeling and regional resource assessment by adhering to a course of nineteenth century determinism. Our critics are very emotional in their charge—their complaints come as shrieks of alarm that suggest that we need to be called back from the precipice. It seems to me that we should expect these warnings as we leave the fold of traditional geologic studies and trek off toward the field of forecasting.

As Tom Ovenshine (1986) pointed out, geologists are called on from beyond our profession to produce mineral-resource assessments in a form amenable to economic analysis and decision making. And this will continue to occur. This idea lurked behind his decisiveness in wanting numerical estimates to start showing up in the conclusions of mineral-resource assessment studies. Knowing that we were expected to produce our assessment of the U.S. Forest Service wilderness areas in the Pacific Mountain system in this form in a short period of time forced us to solve many of the problems that separated the determinism of the geologist from the statistical empiricism of the forecaster. The path along which we progressed as we determined our numerical estimates is discussed below, and the results are shown in Table 10.2. Perhaps the most revealing episode (described below) dealt with our method of determining the estimate for the expected number of porphyry copper deposits.

Dennis Cox had reviewed all the published reports on the wilderness areas under consideration in the State of Washington and had identified those where the geology appeared not only permissive but favorable for the occurrence of this type of deposit. He concluded from his review, particularly from rock alteration data, that 11 porphyry systems had been active within the boundaries of these wilderness areas. He initially told the assessment team that this must mean that there were 11 porphyry copper deposits awaiting discovery in these areas.

Dennis supported his position by rattling off raw data and facts intertwined with implications and conclusions. His discussion included the review of the

Table 10.2. Number of Deposits by Type Considered in the Assessment of Undiscovered Metal Endowment of U.S. Forest Service Wilderness Tracts in the Pacific Mountain System

Deposit type[a]	Metals or minerals contained[b]	Reference for grade-tonnage model[c]	Expected number of undiscovered deposits
Porphyry copper	Cu, Mo, Au, (Ag)	a, p. 21	3.5
Massive sulfide in felsic and intermediate volcanic rocks (Medford type)	Cu, Zn, Au, Ag, (Pb)	c	5.5
Low-sulfide quartz gold (Medford type)	Au	c	7.5
Low-sulfide quartz gold (Sierra Nevada type)	Au, (Ag)	b, p. 54	4.2
Podiform chromite (California type)	Cr	a, p. 3	162.5
Silica carbonate mercury	Hg	b, p. 59	5
Subaerial volcanogenic manganese	Mn, (P)	b, p. 65	1
Hot-springs mercury	Hg	b, p. 62	0.5
Tungsten skarn	W	a, p. 49	2.0
Epithermal gold, quartz-adularia type (Nevada-California type)	Au, Ag	c	2.1
Copper skarn	Cu, (Au, Ag)	a, p. 38	0.1
Synorogenic synvolcanic nickel	Cu, Ni, (Co, Pd, Pt, Au)	b, p. 7	0.1
Zinc-lead skarn	Cu, Zn, Pb, Ag	b, p. 26	1.5
Molybdenum porphyry (low-fluorine type)	Mo	a, p. 31	0.5
Placer gold	Au	None	[d]
Buried Teritiary placers	Au	None	[d]
Massive sulfide, Cyprus type	Cu, (Zn, Pb, Au, Ag)	a, p. 52	None
Sediment-hosted submarine exhalative zinc-lead	Zn, Pb, Ag	a, p. 69	None
Red-bed–green-bed copper	Cu, Ag	None	None
Chrysotile asbestos	Asbestos	b, p. 23	None
Nickel laterite	Ni, (Co)	b, p. 95	None[e]

[a]Descriptive models corresponding to most of these deposit types are in Cox (1983a, b).

[b]Metals in parentheses were not included in the determination of undiscovered metal endowment for this study because they are mostly potential byproducts and grade data were not available for all the deposits used to construct the grade-tonnage models; in the case of subaerial volcanogenic manganese deposits, phosphorus is important as an impurity and not as a resource.

[c]Reference to publication source (and where appropriate, page number) of grade-tonnage model; key: a = Singer and Mosier (1983a), b = Singer and Mosier (1983b), c = Singer et al. (1983).

[d]Number of deposits not estimated. 3

[e]No undiscovered deposits expected although identified deposits occur in the study area.

general geology, exploration geochemistry, and geophysical data. He pointed out that a large porphyry copper deposit that had been discovered in the vicinity of the wilderness areas contained the alteration patterns in question. He also stated that except for the large copper deposits in Arizona, Washington State would have been our main producer of copper today. However, the Arizona copper districts do exist, and they have the additional advantage over those in Washington of being in relatively flat warm places; the mountains of western Washington are rugged, and the climate is cold. In Arizona, you can stop your car beside the road, get out, and walk up to any number of prophyry copper deposits; you cannot do that in the Cascades. As Dennis made this argument, which had mostly to do with relative costs, he swept away our cultural bias that had favored Arizona as a preferred place to predict the occurrence of additional undiscovered porphyry copper deposits.

The discussion then shifted to the other members of the team who, on the basis of their experience gained from working in the mining industry and/or from training in statistical principles, examined Cox's assertion about the 11 porphyry copper deposits. They argued that the existence of 11 altered areas may be proof that 11 porphyry systems had been active, but it was not proof that 11 porphyry copper deposits of commercial grades and tonnages had been implaced. We were seeking an estimate for the number of deposits that were expected to occur *on* our porphyry copper grade and tonnage model; that is, within that domain of the scatterplot shown in Fig. 10.5. This domain also is represented schematically as domain B in Fig. 10.9.

As is commonly done in these discussions, an example from a game of chance was used to illustrate the meaning of the concept of a mathematical expectation. Although an image of the statistical properties of the distribution of the number of undiscovered mineral deposits occurring in a region can be made in many ways, the relation among the parameters (expectations) that control the outcome of the cards drawn in poker hands or the roll of a pair of dice commonly are used for that purpose. In a fair poker hand, for example, one out of every four cards dealt will be a heart; there is some probability that no hearts will occur, a large probability that two hearts will occur, and a nontrivial probability that all five cards in a fair hand will be hearts. By making the analogy between the variation of such pure-chance phenomena and the uncertainty associated with the occurrence of undiscovered mineral deposits, we would say that if we expected 11 undiscovered porphyry deposits to occur, we are also saying that there is a reasonable chance that 10 or 9 deposits could exist. The implication is that there is also some chance that five deposits exist, and, although it could be diminishingly small, there is some chance that no undiscovered porphyry deposits exist when evidence has been found that 11 porphyry systems have been active.

Normally, in dealing with uncertain events, an expectation that 11 porphyry deposits occurred would mean that there was a probability that less than 11 deposits could exist and also that more than 11 could exist. Dennis objected immediately when it was pointed out that an expectation of 11 undiscovered deposits meant that there is a probability for the existence of 15 or maybe even 20 porphyry deposits. His counterpoint was that, relative to the size of the wilderness areas under consideration, this type of deposit is so large and has such obvious geochemical alteration patterns, geophysical signatures, and geologic settings that the occurrence or nonoccurrence of this type of deposit would be noticed. The collective wisdom of the team did not disagree with this statement. Dennis temporarily held the idea that 11 porphyry deposits existed under the 11 sites he had identified.

Those of us who were familiar with the rate of success in drilling exploration prospects for porphyry copper deposits stated that, after drilling, only one out of two or three thoroughly worked-up porphyry-type prospects will prove to have a commercial grade and tonnage. We also asserted that one out of two porphyry systems was barren and contained only iron pyrite in the sulfide phase. Dennis used the common defense of asking for the documentation of these ratios in the published record. He knew that he had a point because little such success ratio data existed in the literature. He also knew that we could produce a convincing argument from our industrial experience and current contacts.

Reasoning that the 11 sites he had identified were equivalent to drillable prospects and knowing that the discovery success ratio had to be less than 0.5 led us quickly to conclude that the expected number of undiscovered deposits had to be five or fewer. After further debate over the meaning of the data, we agreed that the expected number of porphyry copper deposits under the 11 prospects was less than 4 and more than 3. As best as I can tell, the only time in the assessment process when fatigue had any effect on the outcome of the estimation of an expectation was when we decided to split the difference and produced an estimate of three and one-half undiscovered porphyry copper deposits (Table 10.2).

Toward the end of the above-mentioned process, Dennis asked a question that sent those of us who were trained in statistical analysis scrambling for an answer. He wanted us to explain what we meant when we said that our estimate of the expected number of undiscovered deposits had to be *on* the porphyry copper grade and tonnage model. Dennis had no interest in a squabble. He just wanted to know what it meant to be *on* a histogram. I think he saw it as a concept that we believed in strongly and wondered why it was so difficult for us to explain it to him. From the way he stated his question, he must have suspected that this idea was subtle and required statistical training to fathom. He asked, "Does being *on* the model mean that the undiscovered deposits are just like the average deposit discovered in the past?" When we said that it did not, he asked what it did mean. We got as far as saying that it meant being *on* the model in the sense of probability and that it was like being inside the space covered by all the grade and tonnage points. Had Dennis probed further, he would have discerned that most of any empirical grade and tonnage model is empty space, never to be "filled out" by points representing additional discoveries. What, then, could it mean to be on something that was almost all holes and was always going to be mostly nothing? To the determinist mindset, the ideas that are used in applying probability theory to historical data to make predictions about the outcome of future events may well take an extra effort to understand. Dennis trusted us as one trusts one's surgeon during a preoperative discussion. His questioning was an effort to determine as best he could what was going to happen next.

With only one exception, our estimates for the expected number of undiscovered deposits of each of the other mineral deposit types were achieved by using the same subjective procedure as was used to estimate the expectation for the number of porphyry copper deposits. For several of the deposit types, we were able to estimate the number of undiscovered deposits expected to occur after only modest debate. In each of these cases, a single geologic variable usually determined the size of the estimate; for example, the presence or absence of chert beds within the otherwise permissive volcanic terrains of several wilderness areas was the main geologic criterion used to estimate the expected number of undiscovered volcanogenic manganese deposits.

The estimation of the expected number of undiscovered synorogenic-synvolcanic nickel deposits involved the belief that several of the wilderness areas were located where the regional geology was permissive for this type of deposit to occur; however, no site-specific data could be produced to make a numerical estimate. The group agreed that the areas in question could not be totally disregarded because we were unable to acquire such data. We decided to report a small expectation for deposit occurrence, but we could not decide on how close to zero it should be. The device used to determine the expectation that one-tenth of an undiscovered deposit of this type would occur was to pose the question of how many areas we would need to have before we could be comfortable with the assertion that one of this type of deposit would be expected to occur. We determined that 10 such areas were needed. The resulting expectation was then set at one-tenth of a deposit.

The only departure that we made from the use of the subjective assessment format was when we used a linear regression procedure to estimate the number of undiscovered podiform chromite deposits expected to occur in the wilderness areas. The estimate of 162.5 such deposits remaining to be discovered was determined from a regression equation that uses a measurement of the surface area of serpentinite outcrop as the independent variable (Page and Johnson, 1977). This type of chromite deposit has been judged to occur with uniform spatial density in permissive sepentinite terrains.

The estimates of the expected number of undiscovered mineral deposits (Table 10.2) were used as input data to the computer simulation program (Fig. 10.11). This program summarized the undiscovered endowment in terms of the total quantities of the undiscovered metal commodities. This computer simulator contained many of the common structural elements used in these

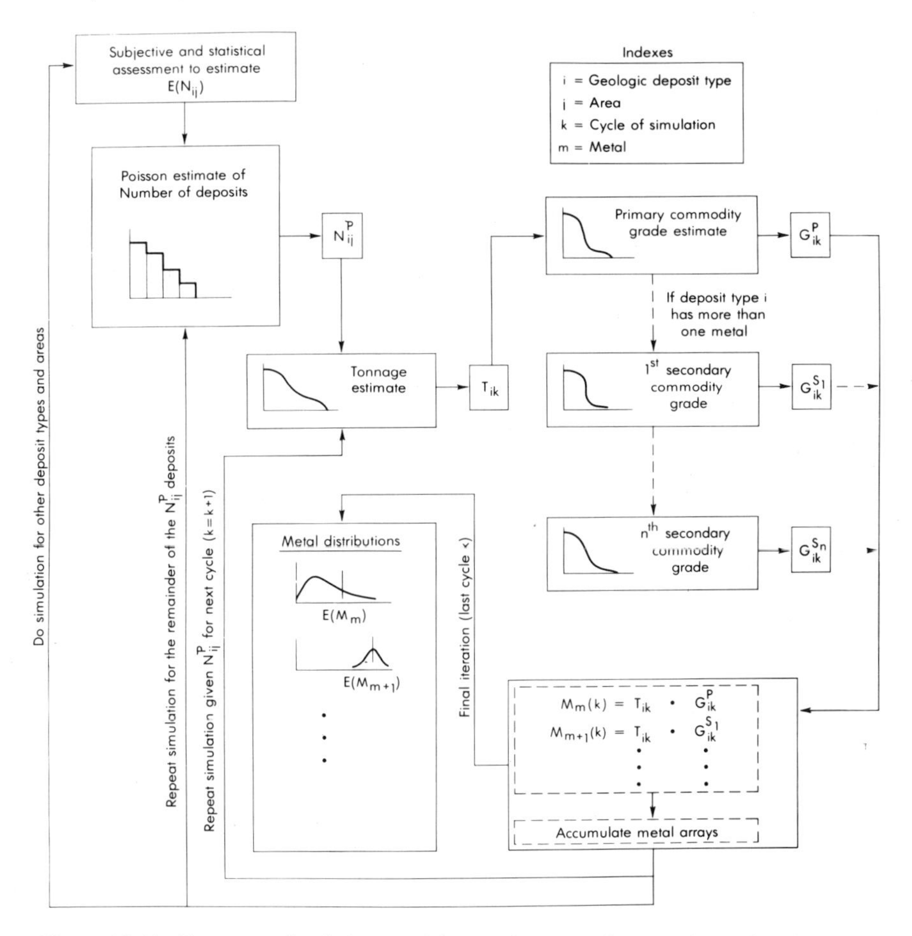

Figure 10.11. Computer simulation model to estimate undiscovered metal endowment. Notes: $E(N_{ij})$ = expected number of deposits of type i in area j, N_{ij}^{P} = number of deposits of type i in area j computed from a Poisson distribution, T_{ik} = tonnage for model type i on cycle k, G_{ik}^{P} = grade for the primary commodity in model type i on cycle k, G_{ik}^{S1} = grade for the first secondary commodity i model type i on cycle k, $M_m(k)$ = expected quantity of metal m generated on cycle k, and $E(M_m)$ = expected quantity of metal m in all areas.

types of simulators; for example, a Poisson generator was used to unfold the expectations (Table 10.2, column 4) into discrete realizations that, in turn, were used as loop counters in the simulator. The relation between the average grade and the tonnage of the deposits in the various grade and tonnage models was assumed to be independent, except in the case of the epithermal quartz-adularia mineral gold deposit type. In this deposit type, the data supported the hypothesis that an inverse relation exists between the ore tonnage and its average gold grade. More recently, statistically significant correlation structures have been found in a number of grade and tonnage models (Cox and Singer, 1986). The computer simulator used by Root et al. (1986) to summarize the size of the endowment of undiscovered gold resources in the Tonapah, Nevada, two-degree sheet uses one of these more complex correlation structures.

The summary of the overall assessment of the undiscovered mineral endowment of the wilderness areas by metal is displayed in Table 10.3. The large skewness observed in the distributions that characterize the size of the undiscovered mineral endowment by metal caused us to choose the medians of these distributions as the most meaningful measure of the central tendency in these distributions. The arithmetic mean is presented for comparison. To give an impression of the relative magnitude of the undiscovered mineral resources occurring in these wilderness areas, an endowment-to-consumption ratio was computed (Table 10.3, column 6).

A policy analyst trying to evaluate the impact of wilderness land withdrawals in the Pacific Mountain system on the mining sector of the economy can determine a value (a social cost) for the withdrawal decision by placing the data in column 6 of Table 10.3 into his policy objective function; for example, by assuming that the undiscovered gold-bearing deposits occurring in wilderness areas can be discovered and put into production within a relevant time frame, the consequence of the withdrawal decision is to remove from the economy an amount of gold worth 1.2 years of domestic gold consumption. In reality, these 1.2 years of consumption would be spread over many years of consumption. Depending on the structure of the cost and the benefit trade-offs in the policy objective function, the magnitude of this loss in domestic gold production may or may not cause the analyst to worry about the decision. Similar calculations

Table 10.3. Estimates of Undiscovered Metal Endowment of 91 U.S. Forest Service Wilderness Tracts in the Pacific Mountain System (in thousand metric tons of metal or oxide)

Metal or oxide	Median undiscovered metal endowment	Upper and lower deciles		Mean endowment	E/C[a] ratio
		90%	10%		
Chromic oxide (Cr_2O_3)	60	40	100	68	0.1
Copper	5,700	850	23,000	10,000	2.6
Gold	0.15	0.038	0.63	0.30	1.2
Lead	110	0	1,700	880	0.1
Manganese	7.6	0	350	180	0.01
Mercury	2.7	0.27	20	11	1.3
Molybdenum	220	25	1,100	460	8.5
Nickel	0	0	0	7.0	0
Silver	1.6	0.25	9.2	4.5	0.4
Tungsten (WO_3)	23	0	400	220	2.0
Zinc	640	89	3,400	1,700	0.7

[a]Ratio of median undiscovered metal endowment (column 2) to average annual U.S. apparent consumption for the years 1979–1982 (from U.S. Bureau of Mines, 1984).

can be made for each of the other metals and also for the aggregate undiscovered mineral-resource endowment. The removal of the quantities of metal shown in Table 10.3 should be evaluated across the time frame in which the availability of the different metals move through their abundance-scarcity cycles before a comprehensive meaning is given to a withdrawal decision.

When we completed our assessment report (Drew et al., 1986), which featured Table 10.3 as its centerpiece, I presented to Tom Ovenshine the first copy of the report. He quickly looked through it and found the table. As he ran his finger down columns 2 and 6, he observed, "There do not seem to be very many undiscovered mineral resources in this batch of wilderness areas!" I never asked if this statement was positive or negative. As I recollect, Tom looked at me with an unfathomable expression and shook the report in his hand and said something like, "This is just fine." I concluded that, in the same sense as the stock market moves to discount good and bad news before it happens, Tom had discounted the report's existence well before it was written. A year or so later, he would remember that he had ordered up this project and would mention the assessment and its outcome with some satisfaction. In a quiet tone, he also noted that it was the result of a pact that we had signed in Marquette.

Glenn Allcott replaced Ovenshine as chief of the Office of Mineral Resources in 1984. In time, Glenn would issue his own opinion of what we had done. His summary, which ran more toward the positive side, stated that producing mineral-resource assessments by estimating the number of undiscovered deposits by deposit type was a step in the right direction. He attached significance to the fact that this format allowed the elements in the assessment to be added up by anyone in any way they liked. He also noticed that this format was conducive to communicating the results of an assessment to the nongeologist. As I listened to these words, my reflex reaction was to put my head down. My thoughts ran to that snowy morning several years before in Marquette, where I learned for sure that there is an assignment right behind an office chief's praise.

Epilogue

My colleagues and I are continuing to study the phenomenon of the economic truncation of observed oil and gas field size distributions. These studies are providing additional data and are furthering our understanding of the ideas presented in Chapters 8 and 9. Of these studies, two were performed in the federal and state waters of the Texas offshore (Drew et al., 1988b), and one was performed in the Frio Strandplain play in onshore Texas (Drew et al., 1988a). They have enabled us to obtain data that demonstrate the nature of the truncation effect across a cost barrier in the same offshore operational environment and, by comparison, how this effect is manifested in observed distribution of discoveries in a directly adjacent onshore exploration play (Fig. E.1). The data from the Frio Strandplain play also unequivocally demonstrate the nature of the shifting to the left of the density mass of oil and gas fields as the exploration process gradually exhausts the parent population of oil and gas fields. These data have been useful in further justifying the application of discovery process models in the assessment of petroleum resources in partially explored areas.

To isolate the effect of cost truncation within this region, the offshore region shown in Fig. E.1 was partitioned into two segments, which share a common petroleum geology and operating conditions for petroleum exploration. The data used in this study were the complete exploration history through 1983 for the state and federal waters out to 200 meters water depth. During this period, 221 oil and gas fields were discovered—95 fields in the state waters and 126 fields in the federal waters. The cost of exploring and developing a field of the

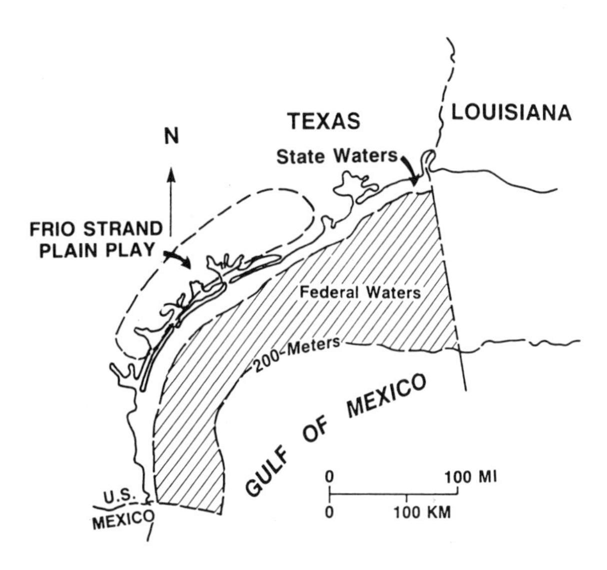

Figure E.1. Location of the onshore Texas (Frio Strandplain exploration play) and two offshore Texas (state and federal waters) exploration areas.

same size in the federal waters is about twice that in the state waters. Most of this difference is attributable to the deeper depth of water on the federal acreage. The economic force that caused these fields to be discovered where they were was the large increase in the price of natural gas that occurred during the middle 1970s and early 1980s (Fig. E.2).

The effect of the different levels of cost truncation in the state and federal waters is displayed in the observed field size distributions shown in Fig. E.3. The distribution for the federal waters is truncated in field size class 11 (3.04 million to 6.07 million barrels of oil equivalent), whereas the distribution for the state waters is in field size class 9 (0.76 million to 1.52 million barrels of oil equivalent). Doubling the costs of exploration and development has had the effect of shifting the point of truncation in the observed field distribution outward by two field size classes. Note that the truncation of these two distributions is very distinct, and, as time passed and discoveries were made, the observed frequency distributions of discoveries built upward and toward the left against these cost barriers. Data such as this have been useful in furthering our understanding of the range of effects that cost truncation can have on the observed distribution of discoveries. In a few words, by making comparisons using the data shown in Fig. E3, we learned how to unmask the effects of the economic overprint placed on an observed field size distribution by the economic conditions that prevail in any particular region. Unmasking this effect is a most important element in the scheme that we use to estimate the number

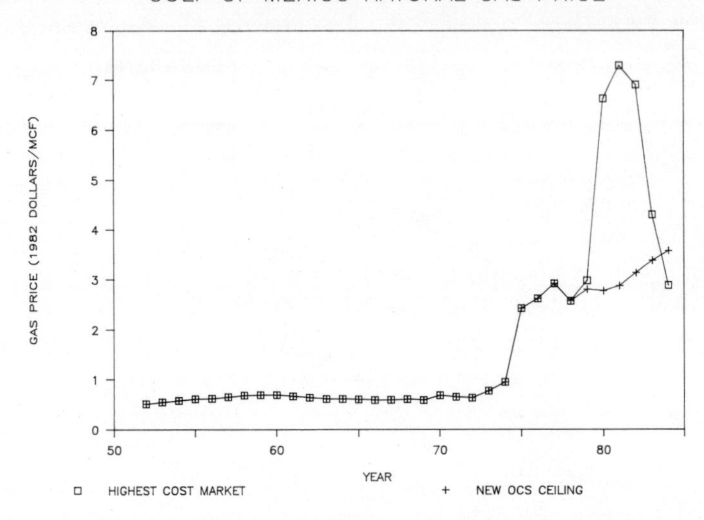

Figure E.2. Estimated prices in 1982 dollars per thousand cubic feet of newly discovered natural gas found in the Gulf of Mexico. The post-1977 prices are represented by two curves. The higher cost market curve represents gas priced in a deregulated market because of its greater production costs due to drilling depth or reservoir pressure. These data are from national averages. The other curve shows regulated price ceilings for newly discovered outer continental shelf gas as set by the Natural Gas Policy Act of 1977.

of oil and gas fields remaining to be discovered that have sizes either above or below the economic truncation point.

In addition to exhibiting distinct evidence of cost truncation, the observed field size distributions show the leftward shifting of the density mass of discoveries with the application of exploration effort over time. Note the consistent leftward shift of the density masses and the modal size class in the state and federal waters over time in Fig. E.3. The small number of discoveries made to the left of the truncation points in these distributions were for oil and gas fields that either had sufficient flow rates to justify profitable development or were developed to recover part of the sunk costs. In a qualitative sense, we can say that without any analysis other than the inspection of the observed size distribution, there are some 70–80 fields in field size classes 8–10 remaining to be discovered in the federal waters after 1983. Similarly, in the state waters, there are 40–50 oil and gas fields in field size classes 6–8 remaining to be discovered after 1983.

The observed field size distribution for onshore exploration plays are truncated in a somewhat different manner than for offshore exploration plays. The principal difference is that the distribution has a much more diffuse character in the vicinity of the modal size class than in offshore regions. To illustrate this fact, compare the two distributions shown in Fig. E.3 that have well-defined

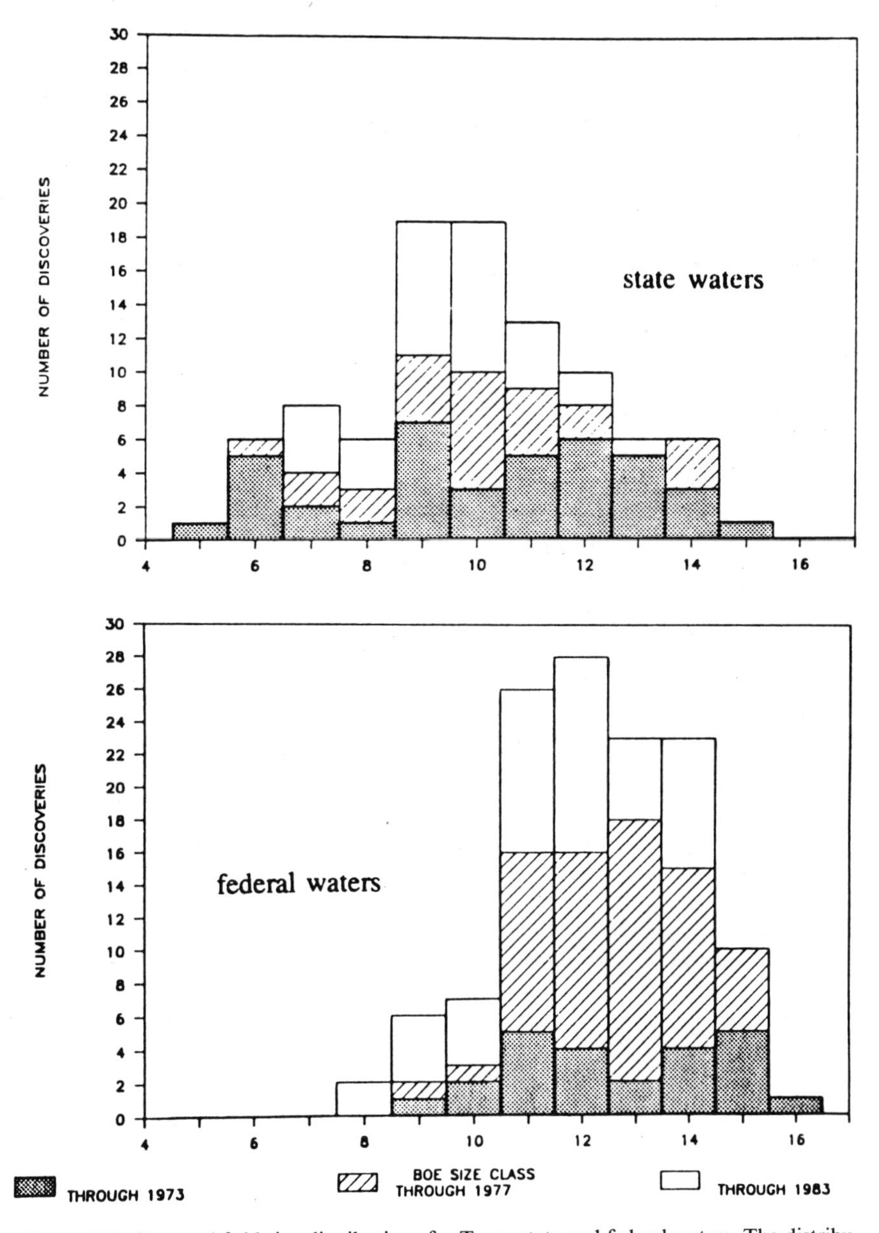

Figure E.3. Observed field size distributions for Texas state and federal waters. The distribution for state waters is truncated in field size class 9, and for federal waters in class 11. BOE is barrels of oil equivalent.

truncation points (size classes 11 and 9) with the much more diffuse character displayed in the vicinity of the mode in the observed distribution for the Frio Strandplain play (Fig. E.4). In none of the three time segments for which this distribution has been constructed is there a sharp break in or near the modal size class of the distribution as there is in the distributions for the adjacent state and federal water offshore regions. The sharp breaks in the distributions for the offshore regions are a consequence of the large incremental costs associated with the installation of offshore production facilities. In onshore exploration plays, such as the Frio Strandplain play, no such sharp breaks in the observed distribution occur because the decision to develop a discovery is weighed against a continuum of drilling and development costs. Truncation in the observed field size distribution of such regions clearly exists, and, as was demonstrated in Chapters 8 and 9, it occurs in the vicinity of the modal size class of the observed size distribution.

The study of the history of the Frio Strandplain play has provided unequivocal evidence for the systematic shifting of the size frequency of discoveries to the left as the exploration process exhausts the parent population of oil and gas fields during an exploration play. This systematic leftward shift in density mass through time provides the data that discovery process modelers need to pursue confidently the activity of forecasting the size distribution of the oil and gas fields remaining to be discovered in a partially explored area and the rate at

Figure E.4 Observed field size distribution for the Frio Strandplain play, onshore Texas. The distribution has three segments representing the cumulative number of fields discovered through the specified year.

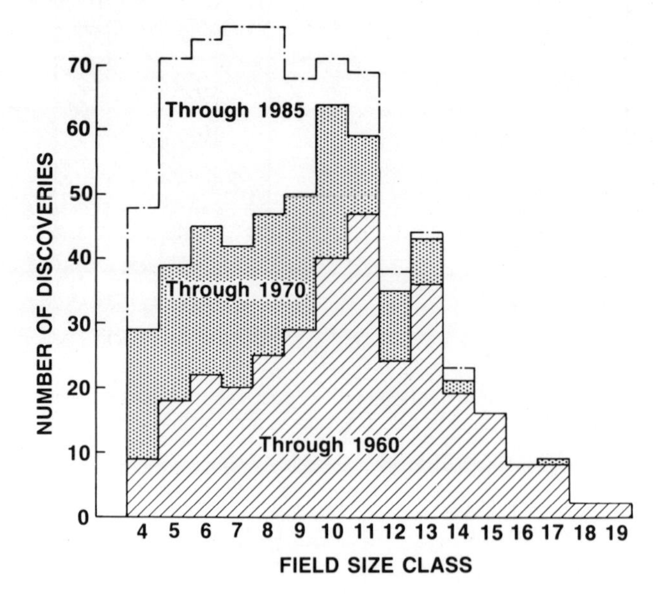

which they will be discovered in the future as a function of the number of wildcat wells drilled.

The data from the Frio Strandplain play also have been useful to us in other ways. They have provided us with our best example of the progressive behavior within a very large play. In this play, over 800 oil and gas fields that collectively contained more that 10 billion barrels of oil equivalent were discovered by 1985. This play also provides data on what happens as an exploration play progresses to an advanced stage of physical exhaustion of the parent population of oil and gas fields. In this regard, note that although 295 fields were discovered between 1971 and 1985, more than 85 percent (253 fields) contained less than 0.729 barrel of producible oil equivalent. Further, the discovery history of this play shows that the modal size class in and of itself is an important frame of reference. I have stressed its position as the point in the observed size distribution where economic truncation occurs and as a necessary element in the application of discovery process models. In addition to the position of the modal class without regard for the formal use of discovery process models, it also contains substantial information that is useful for making qualitative conclusions about the sizes of fields that remain to be discovered in a partially explored region. Note that the modal size class in the size distribution of discoveries in the Frio Strandplain play in 1960 was in size class 11. Next note that of the 467 oil and gas fields that were discovered between 1961 and the end of 1985, only 30 were larger in size than the mode that had been established in 1960 for this exploration play. These data support the conclusion that the majority of the oil and gas left to be discovered in these areas had to be in fields that were smaller in size than the modal size class of the observed field size distribution. As described in Chapter 7, we arrived at this conclusion from the application of discovery process models in such areas as the Miocene-Pliocene and the Pleistocene trends in the offshore Gulf of Mexico. As I mentioned in that discussion, some of our colleagues expressed concern as to the validity of our conclusion. It is data from such highly advanced plays as the Frio Strandplain that add credence to such conclusions. I believe that the modal size class in an observed field size distribution for a partially explored area can be used to state that the amount of oil and gas remaining to be discovered in that region mostly occurs in fields that are smaller than the size of those in the modal size class. Of course, exceptions will be found, but, as a rule of thumb, I believe that this conclusion will be proved to be valid as the exploration process continues to be applied to the petroliferous regions of the world.

References

Agterberg, F. P., Chung, C. F., Fabbri, A. G., Kelly, A. M., and Springer, J. S. 1972. Geomathematical evaluation of copper and zinc potential of the Abitibi area, Ontario and Quebec. Geological Survey of Canada Paper 71–41, 83.

Arps, J. J., and Roberts, T. G. 1958. Economics of drilling for Cretaceous oil on east flank of Denver-Julesburg basin. American Association of Petroleum Geologists Bulletin 42(11), 2549–66.

Attanasi, E. D., and Drew, L. J. 1985. Lognormal field size distributions as a consequence of economic truncation. Mathematical Geology, 17(4), 335–51.

Attanasi, E. D., and Haynes, J. L. 1983a. Economics and the appraisal of conventional oil and gas resources in the western Gulf of Mexico. Society of Petroleum Engineers of AIME, SPE Preprint 11287, Dallas, TX, March 3–4.

Attanasi, E. D., and Haynes, J. L. 1983b. Future supply of oil and gas from the Gulf of Mexico. U.S. Geological Survey Professional Paper 1294.

Attanasi, E. D., and Haynes, J. L. 1984. Economics and appraisal of conventional oil and gas in the western Gulf of Mexico. Journal of Petroleum Technology, 36(12), 2171–80.

Baker, R. A., Gehman, H. M., James, W. R., and White, D. A. 1984. Geologic field number and size assessments of oil and gas plays. AAPG Bulletin, 68(4), 426–37.

Barouch, E., and Kaufman, G. M. 1977. Estimation of undiscovered oil and gas. In Proceedings of the Symposia in Applied Mathematics, 21 (Providence, RI, American Mathematical Society, 1977), 77–91.

Barton, P. B., Jr. 1986. User-friendly mineral deposit models. In Cargill, S. M., and Green, S. B. (eds.) Prospects for mineral resource assessments on public lands: Proceedings of the Leesburg workshop. U.S. Geological Survey Circular 980, 94–110.

Bloomfield, P., Deffeys, K. S., Watson, G. S., Benjamini, Y., and Stine, R. A. 1979. Volume and area of oil fields and their impact on order of discovery. Princeton University, Department of Statistics Technical Report, Princeton, NJ, 53.

Bois, C. 1975. Petroleum zone concept and the similarity analysis—Contributions to resource appraisal. In Studies in Geology, No. 1—Methods of estimating the volume of undiscovered oil and gas resources. American Association of Petroleum Geologists, Tulsa, OK, 87–89.

Constantinou, George, and Govett, G. J. S. 1973. Geology, geochemistry, and genesis of Cyprus sulfide deposits. Economic Geology, 68(6), 843–58.

Cox, D. P., and Singer, D. A. 1986. Mineral deposit models. U.S. Geological Survey Bulletin 1693, 378.

Dolton, G. L., Carlson, H. H., Charpentier, R. R., Coury, A. B., Crovelli, R. A., Frezon, S. E., Khan, A. S., Lister, J. H., McMullin, R. H., Pike, R. S., Powers, R. B., Scott, E. W., and Varnes, K. L. 1981. Estimates of undiscovered recoverable conventional resources of oil and gas in the United States. U.S. Geological Survey Circular 860.

Drew, L. J. 1974. Estimation of exploration success and the effects of resource base exhaustion via a simulation model. U.S. Geological Survey Bulletin 1328, March 1974.

Drew, L. J. 1975a. Linkage effects between deposit discovery and post-discovery exploratory drilling. Journal Research, USGS, 3(2), 169–79.

Drew, L. J. 1975b. Analysis of a petroleum deposit success ratio curve. Mathematical Geology, 7(5/6).

Drew, L. J. 1980. Firm size and performance in the search for petroleum, unpublished M.S. thesis. Virginia Polytechnic Inst., Blacksburg, VA, Sept. 1980.

Drew, L. J., and Attanasi, E. D. 1980. Firm size and performance in the search for petroleum. Dept. of Energy Symposium on Oil and Gas Supply Modeling, June 18–20, NBS Special Pub. 631, 1982, 466–89.

Drew, L. J., Attanasi, E. D., and Schuenemeyer, J. H. 1988b. Observed oil and gas field size distribution: A consequence of the discovery process and the prices of oil and gas. Mathematical Geology, 20(8), 939–53.

Drew, L. J., Bliss, J. D., Bowen, R. W., Bridge, N. J., Cox, D. P., DeYoung, J. H., Jr., Houghton, J. C., Ludington, Steve, Menzie, W. D., Page, N. J., Root, D. H., and Singer, D. A. 1986. Quantification of undiscovered mineral-resource assessment—the case study of U.S. Forest Service wilderness. Economic Geology, 81, 80–88. (Also published as U.S. Geological Survey Open-File Report 84-658 in 1984.)

Drew, L. J., and Campbell, T. J. 1970. An economic analysis of the "Inexco Deal" and additional variations. Cities Service Exploration and Production Research Report G70-11.

Drew, L. J., Grender, G. C., and Turner, R. M. 1983. Atlas of discovery rate profiles showing oil and gas discovery rates by geologic province in the United States. U.S. Geological Survey Open-File Report 83-75.

Drew, L. J., and Root, D. H. 1982. Statistical estimates of tomorrow's offshore oil and gas fields. Ocean Industry, 17(5), 54–66.

Drew, L. J., Schuenemeyer, J. H., and Attanasi, E. D. 1988a. A rationale for estimating the number of small undiscovered oil and gas fields. U.S. Geological Survey Circular 1025, McKelvey Forum, Feb. 29–March 2, 1988, Denver.

Drew, L. J., Schuenemeyer, J. H., and Bawiec, W. J. 1979. Physical exhaustion maps of the Cretaceous D-J sandstone stratigraphic interval in the Denver basin. U.S. Geological Survey I—Map no. 1138.

Drew, L. J., Schuenemeyer, J. H., and Bawiec, W. J. 1982. Estimation of the future rates of oil and gas discoveries in the Gulf of Mexico. U.S. Geological Survey Professional Paper 1252.

Drew, L. J., Schuenemeyer, J. H., Mast, R., and Dolton, G. 1987. Estimates of the ultimate number of oil and gas fields expected to be found in the Minnelusa play of the Powder River basin. U.S. Geological Survey Open-File Report 87-443.

Drew, L. J., Schuenemeyer, J. H., and Root, D. H. 1978. The use of a discovery-process model based upon the concept of the area of influence of a drill hole to predict petroleum discovery rates in the Denver basin. Rocky Mountain Association of Petroleum Geologists guidebook, 31–34.

Drew, L. J., Schuenemeyer, J. H., and Root, D. H. 1980. Resource appraisal and discovery rate forecasting in partially explored regions: Part A—An application to the Denver basin. U.S. Geological Survey Professional Paper 1138.

Duke, M. A., and Hutchinson, R. W. 1974. Geological relationships between massive sulfide bodies and ophiolitic volcanic rocks near York Harbour, Newfoundland. Canadian Journal of Earth Science, 11, 53–69.

Eckstrand, O. R., (ed.) 1984. Canadian mineral deposit types: A synopsis. Geological Survey of Canada Economic Geology Report 36.

EIA (Energy Information Agency). 1983. An economic analysis of natural gas resources and supply. Working Draft A, DOE/EIA-0388, March 1983.

Erickson, R. L., compiler. 1982. Characteristics of mineral deposit occurrences. U.S. Geological Survey Open-File Report 82-795.

Farmer, R. D. 1982. Outer continental shelf (OCS) oil and gas supply model. Volume 1—Model summary and methodology description. U.S. Energy Information Administration, Office of Oil and Gas, DOE/EIA-0372/1.

Farmer, R. D., and Zaffarano, R. F. 1982 [1983]. Outer continental shelf (OCS) oil and gas supply model. Volume 2—Data description. U.S. Energy Information Administration Office of Oil and Gas [Report], DOE/EIA-0372/2.

Forman, D. J., and Hinde, A. L. 1985. Improved statistical method of assessment of undiscovered petroleum resources. American Association of Petroleum Geologists, 69(1), 106–18.

Franklin, J. M., Sangster, D. M., and Lydon, J. W. 1981. Volcanic-associated massive sulfide deposits. In Skinner, B. J. (ed.), Economic Geology Seventy-fifth Anniversary Volume. Economic Geology Publishing Company, 485–627.

Gehman, H. M., Baker, R. A., and White, D. A. 1975. Prospect risk analysis. In Davis, J. C., Doveton, J. H., and Harbaugh, J. W., conveners, Probability methods in oil exploration. AAPG Research Symposium Notes, Stanford University, 16–20.

Gehman, H. M., Baker, R. A., and White, D. A. 1980. Assessment methodology, an industry viewpoint. In Assessment of undiscovered oil and gas, Bangkok. United Nations ESCAP, CCOP Technical Publication 10, 113–121.

Gess, G., and Bois, C. 1977. Study of petroleum zones—A contribution to the appraisal of hydrocarbon resources. In Meyer, R. F. (ed.), The future supply of nature-made petroleum and gas. Pergamon Press, 155–78.

Grender, G. C., Rapoport, L. A., and Vinkovetsky, Y. 1978. Analysis of oil-field distribution for sedimentary basins of United States (abs.). American Association of Petroleum Geologists Bulletin, 62, 518.

Griffiths, J. C. 1966. Exploration for natural resources. Journal of Operations Research Society of America, 14(2), 189–209.

Griffiths, J. C. 1967. Mathematical exploration strategy and decision making. Proceedings, 7th World Petroleum Congress, 11, 87–95.

Harris, D. P. 1973. A subjective probability appraisal of metal endowment of Northern Sonora, Mexico. Economic Geology, 68(2), 222–42.

Harris, D. P. 1977. Conventional crude oil resources of the United States—Recent estimates, methods of estimation and policy considerations. Materials and Society, 1, 263–86.

Harris, D. P. 1984. Mineral resources appraisal. Oxford University Press.

Harris, D. P., Freyman, A. J., and Barry, G. S. 1971. The metholology employed to estimate potential mineral supply of the Canadian northwest—An analysis based upon geologic opinion and systems simulation. Canada Dept. Energy, Mines Resources Information Bulletin MR 105-A.

Hendricks, T. A. 1965. Resources of oil, gas, and natural-gas liquids in the United States and the world. U.S. Geological Survey Circular 522.

Hernfindahl, Orris C. 1969. Natural resource information for economic development. Johns Hopkins University Press.

Hewett, D. F., Callaghan, Eugene, Moore, B. N., Nolan, T. B., Rubey, W. W., and Schaller, W. T. 1936. Mineral resources of the region around Boulder Dam. U.S. Geological Survey Bulletin 871.

Hodges, C. A., and others. 1984. U.S. Geological Survey-INGEOMINAS mineral resource assessment of Colombia. U.S. Geological Survey Open-File Report 84-345, 2 sheets, scale 1:1,000,000.

Hubbert, M. K. 1962. Energy resources: A report to the Committee on Natural Resources of the National Academy of Sciences—National Research Council: Washington, DC, National Academy of Sciences—National Research Council Publication 1000-D. [Reprinted, 1973, National Technical Information Service Report PB-222 401.]

Hubbert, M. K. 1965. National Academy of Sciences report on energy resources: Reply (to discussion by John M. Ryan). American Association of Petroleum Geologists Bulletin, 49(10), 1720–27.

Hubbert, M. K. 1967. Degree of advancement of petrolcum cxploration in United States. American Association of Petroleum Geologists Bulletin, 51(11).

Hubbert, M. K. 1969. Energy resources. In Resources and Man (A study and recommendations by the Committee on Resources and Man. National Academy of Sciences-National Research Council), Chap. 8, 157–242. W. H. Freeman (for National Academy of Sciences).

Hubbert, M. K. 1974. U.S. energy resources, a review as of 1972. Part I–Background paper prepared at the request of Henry A. Jackson, Chairman, Committee on Interior and Insular Affairs, U.S. Senate, 93rd Congress, 2nd Session, Committee Print Serial No. 93-40 (92-75). U.S. Government Printing Office, Washington, D.C.

Hubbert, M. K. 1982. Techniques of predictions as applied to the production of oil and gas, Dept. of Energy Symposium on oil and gas supply modeling, June 18–20, 1980. NBS Special Publication 631, 1982, 16–141.

Kaufman, G. M. 1962. Statistical decision and related techniques in oil and gas exploration. Prentice-Hall.

Kaufman, G. M. 1983. Oil and gas—Estimation of undiscovered resources. In Adelman, Houghton, Kaufman, and Zimmerman, (eds.), Energy resources in an uncertain future. Ballinger.

Kaufman, G. M., Balcer, Y., and Kruyt, D. 1975. A probabilistic model of oil and gas discovery. In Studies in Geology, No. 1-Methods of estimating the volume of undiscovered oil and gas resources. American Association of Petroleum Geologists, Tulsa, OK, 113–42.

Kaufman, G. M., and Wang, J. 1980. Model mis-specification and the Princeton study of volume and area of oil fields and their impact on the order of discovery. MIT Energy Laboratory Working Paper no. MIT-EL 80-003WP. Cambridge, MA.

Koski, R. A., and Derkey, R. E. 1981. Massive sulfide deposits in ocean-crust and island-arc terranes in southwestern Oregon. Oregon Geology, 43(9), 119–25.

Lasky, S. G. 1947. The search for concentrated deposits—A reorientation of philosophy. Technical Publication 2146, American Institute of Mining and Metallurgical Engineering.

Lasky, S. G. 1950a. Mineral-resource appraisal by the U.S. Geological Survey. Colorado School of Mines Quarterly, 45(1A), 1–27.

Lasky, S. G. 1950b. How tonnage and grade relations help predict ore reserves. Engineering and Mining Journal, 151(4), 81–5.

Lafayette Geological Society. 1973. Offshore Louisiana oil and gas fields.

Lee, P. J., and Wang, P. C. C. 1983. Probabilistic formulation of a method for the evaluation of petroleum resources. Mathematical Geology, 15(1), 163–81.

Lindgren, Waldemar. 1933. Mineral deposits (4th edition). McGraw-Hill.

Ludington, S. D. 1984. Preliminary mineral resource assessment of the proposed Mt. Wrightson Wilderness, Santa Cruz and Pima Counties, Arizona. U.S. Geological Survey Open-File Report 84-294.

Mallory, W. W. 1975. Accelerated national gas resource appraisal (ANOGRE). In Haun, J. D., (ed.) Methods of estimating the volume of undiscovered oil and gas resources. American Association of Petroleum Geologists, Tulsa, OK, 23–30.

Marquardt, D. W. 1963. An algorithm for least-squares estimation of nonlinear parameters. Society for Industrial and Applied Mathematics Journal, 11(2), 431–41.

Marsh, S. P., Kropschot, S. J., and Dickinson, R. G. (eds.) 1984. Wilderness mineral potential: Assessment of mineral-resource potential in U.S. Forest Service lands studied 1964–1984. U.S. Geological Survey Professional Paper 1300, 1 and 2.

Mast, R. F., Dolton, G. L., Crovelli, R. B., Powers, R. B., Charpentier, R. R., Root, D. H., and Attanasi, E. D. 1988. Estimates of undiscovered recoverable oil and gas resources for onshore and state offshore areas of the United States. U.S. Geological Survey Circular 1025, 31–32.

Masters, C. D., Root, D. H., and Dietzman, W. D. 1983. Distribution and qualitative assessment of world crude-oil reserves and resources. U.S. Geological Survey Open-File Report 83-728.

McCrossan, R. G. 1969. An analysis of size frequency distribution of oil and gas reserves of western Canada. Canadian Journal of Earth Science, 6(2), 201–11.

McCrossan, R. G. (ed.) 1973. The future petroleum provinces of Canada—Their geology and potential. Canadian Society of Petroleum Geologists Memoir 1.

McCrossan, R. G., and Porter, J. W. 1973. The geology and petroleum potential of the

Canadian sedimentary basins—A synthesis. In McCrossan, R. C. (ed.), The future petroleum provinces of Canada—Their geology and potential. Canadian Society of Petroleum Geologists Memoir 1, 589–720.

McKelvey, V. E. 1960. Relations of reserves of the elements to their crustal abundance. American Journal of Science 258-A (Bradley volume), 234–41.

McKelvey, V. E. 1972. Mineral resources estimates and public policy. American Journal of Science, 60, 32–40.

McKelvey, V. E., and Duncan, D. C. 1965. United States and world resources of energy. In Symposium on Fuel and Energy Economics, Joint with Division on Chemical, Marketing and Economics, 149th National Meeting. American Chemical Society, Division of Fuel Chemistry, 9(2), 1–17.

McKelvey, V. E., and Masters, C. D. 1984. Undiscovered oil and gas resources—Procedures and problems of estimation. Proceedings of the 27th International Geological Congress, 13, VNU Science, 333–52.

Meisner, J., and Demirmen, F. 1981. The creaming method—a Bayesian procedure to forecast future oil and gas discoveries in mature exploration provinces. Journal Royal Statistical Society A, 144(1), 1–31.

Meyer, R. F. 1970. Geologic provinces code map for computer use. American Association of Petroleum Geologists Bulletin, 54(7), 1301–5.

Meyer, R. F. (ed.) 1974. Geologic provinces of the United States. American Association of Geologists, scale 1:5,000,000 (rev. ed.).

Miller, B. M., Thomsen, H. L., Dolton, G. L., Coury, A. B., Hendricks, T. A., Lennartz, F. E., Powers, R. B., Sable, E. G., and Varnes, K. L. 1975. Geological estimates of undiscovered recoverable oil and gas resources in the United States. U.S. Geological Survey Circular 725.

Mitchell, A. H. G., and Garson, M. S. 1981. Mineral deposits and global tectonic settings. Academic Press.

Moore, D. P., and Drew, L. J. 1979. A mineral exploration simulation model. Proceedings Transaction of Council of Economics of the AIME, New Orleans, 77–82.

Nolan, T. B. 1950. The search for new mining districts. Economic Geology, 45, 601–8.

Office of Technology Assessment. 1987. U.S. oil production—Response to low prices. Office of Technology Assessment, United States Congress Draft Report.

Ovenshine, A. T. 1986. Why mineral deposit models came to be used in regional assessment. In Cargill, S. M., and Green, S. B. (eds.), Prospects for mineral resource assessments on public lands: Proceedings on the Leesburg workshop. U.S. Geological Survey Circular 980.

Page, N. J. and Johnson, M. G. 1977. Chromite resources of podiform chromite deposits and exploration for concealed chromite deposits in the Medford-Coos Bay quadrangles, southwestern Oregon. U.S. Geological Survey Open-File Report 77-656.

Peterson, E. U., and Zantop, H. 1980. The Oxec deposit, Guatemala: An ophiolite copper occurrence. Economic Geology, 75, 1053–65.

Petroleum Information, Inc. 1975. Well history control file. Petroleum Information Inc., magnetic tape.

Pratt, W. E. 1942. Oil in the earth. Kansas University Press.

Procter, R. M., Taylor, G. C., and Wade, J. A., 1984. Oil and natural gas resources of Canada 1983. Geological Survey of Canada Paper 83–31, 59.

Richter, D. H., Singer, D. A., and Cox, D. P. 1975. Mineral resources map of the Nabesna quadrangle, Alaska. U.S. Geological Survey Miscellaneous Field Studies Map MF-655K.

Root, D. H., and Attanasi, E. D. 1980. An analysis of petroleum discovery data and a forecast of the date of peak production. In facts and principles of world petroleum occurrence. Canadian Society of Petroleum Geologists Memoir 6, 363–75.

Root, D. H., and Drew, L. J. 1979. The pattern of petroleum discovery rates. American Scientist, 67(6), 648–52.

Root, D. H., and Drew, L. J. 1981. General principles of the petroleum industry and their application to the U.S.S.R. In Energy in Soviet policy, report to Joint Economic Committee of the 97th Congress, June 11, 1981, 127–39.

Root, D. H., Drew, L. J., and Scott, W. A. 1986. Estimate of gold resources in the Tonopah 1 × 2 degree quadrangle, Tonopah, Nevada 1 × 2 degree quadrangle CUSMAP meeting, Reno, NV, Sept. 26, 1986.

Root, D. H., and Schuenemeyer, J. H. 1980. Mathematical foundations, pt. B. In Petroleum-resource appraisal and discovery rate forecasting in partially explored regions. U. S. Geological Survey Professional Paper 1138 A-C, B1–B9.

Ryan, J. T. 1965. Limitations of statistical methods for predicting petroleum and natural gas reserve and availability. Preprint number SPE 1256, 40th Annual Fall Meeting of the Society of Petroleum Engineers of AIME, Denver, CO.

Ryan, J. T. 1973a. An analysis of crude-oil discovery rate in Alberta. Bulletin of Canadian Petroleum Geology, 21(2), 219–35.

Ryan, J. T. 1973b. An estimate of the conventional crude-oil potential in Alberta. Bulletin of Canadian Petroleum Geology, 21(2), 236–46.

Sawkins, F. J. 1984. Metal deposits in relation to plate tectonics. Springer-Verlag.

Schuenemeyer, J. H., and Attanasi, E. D. 1984. Forecasting rate of hydrocarbon discoveries in a changing economic environment. Marine and Petroleum Geology, 1, 313–18.

Schuenemeyer, J. H., and Drew, L. J. 1983. A procedure to estimate the parent population of the size of oil and gas fields as revealed by a study of economic truncation. Mathematical Geology, 15(1), 145–62.

Searle, D. L., and Panayiotou, A. 1980. Structural implications in the evolution of the Troodos massif, Cyprus. In Panayiotou, A. (ed.), Ophiolites: International Ophiolite Symposium, Cyprus, 1979, 50–60.

Sillitoe, R. H. 1972. A plate tectonic model for the origin of porphyry copper deposits. Economic Geology, 67, 184–97.

Singer, D. A. 1975. Mineral resource models and the Alaskan mineral resource assessment program. In Vogely, W. A. (ed.), Mineral materials modeling: A state-of-the-art review. Johns Hopkins University Press, 370–82.

Singer, D. A. 1984. Mineral resource assessments of large regions—Now and in the future. In Geological Survey of Japan, ed., U.S.-Japan joint seminar on resources in the 1990's; June 1984; pub. Earth Res. Satellite Data Analysis Center, 2, 31–40. (Republished in Chishitau News, 1986, 377, 18–25, in Japanese.)

Singer, D. A., Cox, D. P., and Drew, L. J. 1975. Grade-tonnage relationships among copper deposits. U.S. Geological Survey Professional Paper 907-A.

Singer, D. A., and Drew, L. J. 1976. The area of influence of an exploratory hole. Economic Geology, 71(3), 642–47.

Singer, D. A., Mosier, D. L., and Cox, D. P. 1986. Grade and tonnage of porphyry Cu. In Cox, D. P., and Singer, D. A. (eds.), Mineral deposit models. U.S. Geological Survey Bulletin 1693.

U.S. Geological Survey. 1980. Future supply of oil and gas from the Permian basin of west Texas and southeastern New Mexico, Circular 828.

Vidas, E. H., and Duleep, K. G. 1984. Find-rate methodology and resource base assumptions. Prepared for Gas Research Institute Contract 5082-711-0571, Washington, D.C.

White, D. A. 1980. Assessing oil and gas plays in facies-cycle wedges. American Association of Petroleum Geologists Bulletin, 64, 1158–78.

White, D. A., Garrett, R. W., Jr., Marsh, G. E., Baker, R. A., and Gehman, H. M. 1975. Assessing regional oil and gas potential. In Methods of estimating the volume of undiscovered oil and gas resources. AAPG Studies in Geology 1, 143–59.

White, D. A., and Gehman, H. M. 1979. Methods of estimating oil and gas resources. American Association of Petroleum Geologists Bulletin, 63, 2183–92.

Zapp, A. D. 1962. Future petroleum-producing capacity of the United States. U.S. Geological Survey Bulletin 1142-H.

Index